is a book characterised by a deep sense of humility before
nystery of life and the limitations of human knowledge. It is
one that is generously open to the views of those with whom
author disagrees. At the same time McGrath shows, often
a telling example, how Christianity not only gives us a
pse of the bigger picture, but offers a framework of meaning
ling us to cope with our journey from birth to death. Whilst
ving on his vast learning in both science and philosophy
Grath still manages to convey his argument with great clarity
accessibility. It is a book that will challenge all dogmatists,
her scientific or religious, and which will greatly encourage
e who are tentative and searching.'
    – Lord Harries of Pentregarth, former Bishop of Oxford

his personal, scholarly yet gripping account of the human
h for meaning, Alister McGrath reveals this irenic question
as vital in our own times as in previous ages. In the spirit of
erton, Thoreau and C. S. Lewis, and in dialogue with
stine, Pico, Murdoch, and his neighbourhood nemesis
ins, McGrath takes his breathless readers first to a high
ny-view of the science, religion and philosophy of purpose,
en leads us back down to the road where we must make our
ourneys, the richer for our reading.'
                    – Professor Tom McLeish, FRS,
            Professor of Physics, Durham University

D1079139

Also by Alister McGrath

In the Beginning
Deep Magic, Dragons and Talking Mice
C. S. Lewis: A Life
Inventing the Universe

The Great Mystery

Science, God and the Human Quest for Meaning

Alister McGrath

HODDER

First published in Great Britain in 2017 by Hodder & Stoughton
An Hachette UK company

This paperback first published in 2018

1

Copyright © Alister McGrath, 2017

The right of Alister McGrath to be identified as the Author of the Work has
been asserted by him in accordance with the Copyright, Designs and Patents
Act 1988.

A CIP catalogue record for this title is available from the British Library

ISBN 978 1 473 63433 6
eBook ISBN 978 1 473 63434 3

Typeset in Sabon MT by Hewer Text UK Ltd, Edinburgh
Printed and bound in the UK by Clays Ltd, St Ives plc

Hodder & Stoughton policy is to use papers that are natural, renewable
and recyclable products and made from wood grown in sustainable
forests. The logging and manufacturing processes are expected to
conform to the environmental regulations of the country of origin.

Hodder & Stoughton Ltd
Carmelite House
50 Victoria Embankment
London EC4Y 0DZ

www.hodderfaith.com

In memory of my mother

Contents

PART ONE

Wondering about Ourselves

I

Born to Wonder:
Asking Questions; Hoping for Answers

'What a little vessel of strangeness we are, sailing through
this muffled silence through the autumn dark.'[1]
John Banville

Life is a gift. We never asked to be born. Yet here we are, living in
this strange world of space and time, trying to work out what it's
all about before the darkness closes in and extinguishes us. We are
adrift on a misty grey sea of ignorance, seeking a sun-kissed
island of certainty, on which we might hope to find clear answers
to our deepest and most poignant questions. What is the point of
life? Why are we here? And what is it about us that makes us want
to ask these questions?

This book reflects on what it means to be a human being, at a
time when many are wondering whether we can ever sort out the
muddle and chaos of our world. If human beings are so wonderful, why is the world such a mess? Why do we use wonderful
things for such nasty purposes? Why are we so resistant to facing
up to uncomfortable truths about ourselves? These are hardly
new questions. They bubble up, time and time again – especially
when events challenge our easy-going assumptions about our own
future, or that of the world.

During the 'Roaring Twenties', most Americans were happy to
buy into the genial optimism of the age. Like the stock market,
the world seemed to be heading upwards. Then the bubble burst.
The Wall Street crash triggered a financial crisis in Germany,
which gave Adolf Hitler the political impetus he needed to get
elected. By 1934, Germany had turned Nazi. The unwelcome and
unexpected rise of Fascism triggered unease in many quarters.

Perhaps most importantly, it led to an overdue re-examination of some complacent settled assumptions about human goodness and rationality.[2]

Reinhold Niebuhr – a theologian noted for his criticism of the lazy and unthinking optimism of so much Western thinking – spoke of a pervasive sense of cultural unease and disenchantment in 1942, as the world collapsed into global war. 'We have lived through such centuries of hope, and we are now in such a period of disillusionment.'[3] Yet after an all-too-short period following the Second World War during which we dared to hope for a future that lived up to our past, that world-weariness is on the rise again.

So is it time to look at ourselves again, holding up a mirror so that we can see ourselves as we really are, rather than as we like to think we are? As I grew up in the 1960s, I was conscious of a pervasive if understated sense of optimism that now seems to have ebbed away, like a receding tide. Back in the 1960s, culturally defining and lingering memories of the Second World War helped to confirm the belief that things were getting better, and the hope that they would keep on getting better. Yet that spirit of hope now seems to have faded in the face of economic crashes and political crises, the rise of global terrorism, and the growing threat of climate change. Paradise seems to have been postponed – yet again. Perhaps, as Milan Kundera suggested, our longing for paradise is really an unattainable desire to escape from the limiting condition of being human.[4]

A time of crisis and disenchantment calls out for a fundamental rethinking of who we are, rather than collapsing into cynicism and despair. That's what this book tries to do. It draws on both religion and science – two of the richest and most complex elements of modern culture – to explore human nature, especially our quest for meaning in life. In particular, it tries to address what is perhaps the most unsettling question of all, routinely ignored by so many smug and complacent social commentators: *what is wrong with us?* No single human discipline or research tradition is good enough to give the rich, textured and complicated answer that we need if we are to confront our weaknesses and shape our future, both individually and collectively. But we have to confront them, and work out where we go from here. It's all about understanding ourselves, and this mysterious gift of life that has been entrusted to us.

The Quest for Meaning

So what is life about? As far as we know, we're the only species on earth that asks this question, and dares to hope that we might find an answer. It seems that we are born to wonder, not merely to exist.[5] To wonder is to reflect, to turn over in our minds what is known, to expand our imaginative capacity and to ask what greater truth and beauty might lie behind our world or beyond our settled horizons of vision. We want to know why things take their present forms, and whether they point to something deeper.

The question of the meaning of life used to be seen as making philosophy intensely relevant to life.[6] Yet as the philosopher Susan Wolf noted recently, it is hardly ever asked in philosophical circles nowadays – and then only by naïve young students, whose lack of sophistication causes professional philosophers to cringe with embarrassment.[7] Many now wonder whether academic philosophy has lost touch with the questions that really matter to people, and which brought philosophy into being in the first place. That was the view of Henry David Thoreau (1817–62), in his widely read classic *Walden* (1854). 'There are nowadays professors of philosophy, but not philosophers.'[8]

Sadly, Thoreau's words will probably ring true for all too many readers today. Philosophy seems to have become the study of other philosophers, an exercise in academic introspection and professional self-reference rather than an engagement with the deepest questions of life – questions that are now often dismissed as intellectually incoherent or naïve, because they are so difficult to answer. Yet perhaps Milan Kundera needs to be heeded when he remarked that 'it is questions with no answers that set the limit of human possibilities, describe the boundaries of human existence'.[9] Such questions probe our limits, challenging us to take intellectual risks in transgressing the boundaries of a cold rationalism.

Yet while Wolf ruefully notes that discussion of whether life has any meaning now seems to have been 'banished from philosophy', it most certainly has not been marginalised in the everyday lives of ordinary people, who seek meaning, value and purpose in order to make sense of their lives, and meaningfully inhabit our strange and puzzling world. Professional philosophy has not discredited the validity of trying to find meaning in life; it has just embargoed it.[10]

Happily, there is no shortage of others anxious to engage with this ultimate question, and take it seriously. Psychology – an empirical research discipline which is far more attentive to human needs and concerns than philosophy now seems to be – has highlighted how important the question of meaning is to our wellbeing.[11] Human beings seem to yearn for a 'big picture' which helps us feel that we are part of something greater than ourselves.[12] That's just the way we function as human beings.

To explore this further, let's see how the human quest for meaning in life links up with another fundamental human experience – a sense of wonder at the beauty of our world.

Wonder and the Meaning of Life

From time to time, we find ourselves overwhelmed by a sense of awe or mystery, often when confronted with the beauty or majesty of nature, which seems for a moment to intimate a grander vision of reality, perhaps lying beyond the horizons of our experience. Many experience a sense of wonder and joy at the fact that there is anything at all; others when they are struck by the full significance of the astonishing fact that we are alive, and able to behold this strange world in which we find ourselves. It is as if, for only a moment, a veil is removed and we catch a half-glimpsed sight of a promised land, waiting to be mapped and explored.

G.K. Chesterton spoke of the 'object of the artistic and spiritual life' being to 'dig for this submerged sunrise of wonder'.[13] Captured by this vision, we long to know more.[14] It can become a gateway to science, art, literature and religion[15] – in short, to everything that gives value and meaning to human life. A sense that there is indeed some such big picture becomes a driving force for creative exploration, in whose slipstream arise the great human quests for knowledge and wisdom.[16]

We cannot overlook the power of this sense of wonder to excite the poetic imagination, which throws down the gauntlet to what often turn out to be narrowly dogmatic and excessively cerebral accounts of our world, inviting us to consider that there is more to reality than an impoverished rationalist philosophy might allow.[17] Nor can we fail to recognise the capacity of a sense of 'rapturous amazement' (Albert Einstein) to motivate and empower the natural sciences. Richard Dawkins and I disagree about many questions in life, but we both know and delight in the beauty and vastness of the world around us.[18] Yet the immensity of our universe conquers our minds, and forces us to engage with the universe on its own terms. Why? Because it is too vast for the 'all too limited human mind' (Dawkins) to take it in fully.[19]

As Aristotle pointed out more than two thousand years ago, our experience of wonder serves as an invitation to set out on a journey of discovery of our world, in which our mental horizons are expanded and our eyes opened.[20] The natural sciences are ultimately an act of intellectual homage to our universe, as we try to grasp its mysteries with the tools we have at our disposal. Yet all too soon we find that the conceptual systems we forge as intermediaries for this act of comprehension strain to cope with these overwhelming realities, like old wineskins struggling to cope with new wine. Our sense of wonder expresses both a delight in the grandeur and glory of our universe, and a recognition of the inadequacy of our capacity to take it in fully. As we shall see, science and religion, in their different ways, invite us to raise our eyes from the world of what we see around us, and try to imagine a deeper vision of reality which underlies and explains what we observe.[21]

How Does Science Fit Into This?

I opened this chapter with a quote from the Irish writer John Banville, whose early writings show a clear appreciation of the 'rage for order' that underlies the human quest for meaning. Banville notes how scientists such as Copernicus and Kepler sought to impose order on the world, and then tried to live in accordance with the framework of meaning they believed it disclosed. 'I saw a certain kind of pathetic beauty in their obsessive search for a way to be in the world, in their existentialist search for something that would be authentic.'[22]

Yet the plausibility of that vision faded in the twentieth century, confronted with the fragility and provisionality of human knowledge. The cultural investment in science as a tool of discernment of meaning or value proved to be a misjudgement. As its failure became more widely appreciated after the Second World War, Western culture experienced a transition from 'Cartesian certainty to Wittgensteinian despair', in which the early hope of finding the Enlightenment's Holy Grail, the crystalline clarity of rationalist certainties, gradually gave way to a realisation of the irreducible complexity of the world.[23]

Banville chronicles this slow and seemingly irreversible transition from rational certainty to existential despair and cynicism with a graceful prose that sweetens his bitter diagnosis of our situation. What one generation took to be rational certainties were found by another to be cultural constructions. It is a problem that rationalist writers of the eighteenth and nineteenth centuries ignored or suppressed, hoping that the rhetoric of their 'glib and shallow rationalism' (C.S. Lewis) would distract people from its striking lack of traction on reality.[24] While those rational certainties live on in the curious backwater of the 'New Atheism', everyone else is trying to figure out how to cope with the predicament in which we find ourselves. Not even the sciences can deliver secure answers to the questions we ask about meaning, value and purpose.

Albert Einstein explored this point in a landmark lecture at Princeton Theological Seminary in 1939.[25] Einstein insisted that the natural sciences were outstanding in their sphere of competence. Yet he cautioned that 'the scientific method can teach us nothing else beyond how facts are related to, and conditioned by, each other'. Human beings need more than what a 'purely rational conception of our existence' is able to offer. Yet opening up such fundamental questions of meaning and value does not cause us to lapse into some kind of superstitious irrationality. 'Objective knowledge provides us with powerful instruments for the achievements of certain ends, but the ultimate goal itself and the longing to reach it must come from another source.' For Einstein, the fundamental beliefs which are 'necessary and determinant for our conduct and judgments' cannot be developed or sustained in a 'solid scientific way'. Einstein was emphatic that this was not a criticism of science. It was simply an informed and necessary recognition of its limits.

Einstein's point is echoed in a striking statement of Sir Peter Medawar (1915–87), a leading British biologist who championed the public engagement of science: 'Only humans find their way by a light that illuminates more than the patch of ground they stand on'.[26] Medawar's remarks point to the importance of transcendence in the human quest for meaning – the desire to see ourselves as part of a bigger picture, which goes beyond our immediate needs and concerns. Human beings seem to be driven to find something deeper than what can be found through an examination of the empirical world. There is a large body of research literature which suggests that we cope better with our complex and messy world if we feel that we can discern meaning and value within our own lives, and in the greater order of things around us.[27]

There are, of course, some who argue that science, and science alone, can tell us everything we need to know about the meaning of the universe and life. This position is often known as 'scientism', which is generally understood as 'a totalizing attitude that regards science as the ultimate standard and arbiter of all interesting questions'.[28] Some scientists do indeed think that the part of reality that their methods can engage with constitutes the whole of reality; some philosophers have been unwise enough to try and 'assimilate philosophy to the aims, or at least the manners, of the sciences'.[29]

Yet most of the scientists I know would disagree, holding that science fills in part of the 'big picture' of reality – but only part. We need to draw on other sources of wisdom to enrich the highly focused account of reality that science provides. Science is a reliable source of knowledge about our universe, based on what we experience. Yet there is no good reason to suppose that science can offer a complete account of reality. It clearly needs supplementation. So what other resources might amplify our vision of reality?

How Does Religion Fit Into This?

Science is unquestionably a core resource in the human quest for understanding and wisdom. Yet there is another, routinely dismissed by those who limit reality to what reason and science can prove. The sense of awe and wonder at nature which motivates science also turns out to be a gateway to what we so inadequately describe as 'religion'. For the psychologist William James, religion was basically about 'faith in the existence of an unseen order of some kind in which the riddles of the natural order may be found and explained'.[30] Yet James provides us with nothing more than a helpful starting point for reflection here. After all, one of the great themes of classic Greek philosophy was that there was an *archē*, a fundamental principle of order and coherence within our world and our minds.[31] Though not necessarily expressed in what we might nowadays call religious terms, this same idea lies at the heart of Christianity, as it does of so many other religious traditions.

Susan Wolf rightly notes that religion is now one of the most important sources of meaning and value in our culture.[32] As human beings, we need something that will hold together our minds and our hearts, our reason and experience, and not improperly restrict us to the imaginatively dull and impoverished world of rationalism. As the Cambridge physicist Alexander Wood perceptively observed, 'our first demand of religion' is that it should 'illumine life and make it a whole'.[33]

This point was made with particular clarity by Salman Rushdie in his 1990 Herbert Memorial Lecture at Cambridge University. Down the ages, he argues, religion has met three types of needs which have failed to be satisfied by secular, rationalist materialism.[34] First, it enables us to articulate our sense of awe and wonder, partly by helping us grasp the immensity of life, and partly by affirming that we are special. Second, it provides 'answers to the unanswerable', engaging the deep questions that so often trouble and perplex us. And finally, it offers us a moral framework, within which we can live out the good life. For Rushdie, religion or the 'idea of God' provides us with a 'repository of our awestruck wonderment at life, and an answer to the great questions of existence'. Any attempt to describe or define human beings 'in terms that exclude their spiritual needs' will only end in failure.

This book takes a cue from Rushdie's insight. Any comprehensive and reliable account of humanity has to take into account the innate tendency towards religion or spirituality that seems to be an intrinsic aspect of human nature. This does not validate religion or belief in God as *right*; it does, however, indicate that these are both *natural* and *human*. In other words, they are part of what it means to be human, and must therefore be addressed as an integral aspect of human nature. The recognition of this fact is now widely conceded.

Recognising this fact helps us to make sense of what seems, at least at first sight, to be some remarkable inconsistencies – such as the leading New Atheist writer Sam Harris's interest in Eastern mysticism. Yet Harris, like many other atheist writers, has simply recognised the importance of this aspect of human nature – what we might loosely call the 'quest for the spiritual', whether this is framed in theist or atheist terms.[35] It's part of being human, irrespective of where that quest leads us in terms of our thinking about God or religion.

Religion: An Important Idea – an Unhelpful Term

Many are wondering if we need to find a new word for what we traditionally call 'religion'. Growing academic and cultural interest in the phenomenon of religion has made it painfully clear that there is a real problem of defining what this actually is.[36] The word 'religion' was extracted from classic Roman culture, within which its meaning was as limited as it was clear, and imposed by modern Western scholars upon a variety of human phenomena, thus creating the false impression that the term designated some global or universal phenomena. Individual religions certainly exist, yet the global notion of 'religion' is a social construction generated by a human desire to impose firm conceptual distinctions on a complex world. The phenomenon is real enough; the problem lies in the words we use to describe it.

So how do we get round this problem? The simplest answer lies in rejecting what we might call 'essentialist' theories of religion, such as the simplistic cognitive approach found in the 'New Atheism', which sees religion as a set of unproven beliefs.[37] On this view, there is a global reality called 'religion', and every individual religion is a specific instance or example of this universal, possessing the same essential property or properties. Yet despite their obvious differences and divergences, we can nevertheless discern some features that seem to be common across religious traditions. One of these core themes is the development of a 'big picture' of reality, which provides a framework for the discernment of truth, beauty and goodness. The philosopher Keith Yandell offers a good account of this aspect of religion: 'A religion is a conceptual system that provides an interpretation of the world and the place of human beings in it, bases an account of how life should be lived given that interpretation, and expresses this interpretation and lifestyle in a set of rituals, institutions and practices.'[38]

Now it could easily be objected that such systems of meaning are found beyond the category of religion – for example, in Marxism, or the metaphysically inflated 'universal Darwinism' of Richard Dawkins.[39] As the philosopher Mary Midgley pointed out, this helps us understand why Marxism and Darwinism – the 'two great secular faiths of our day' – display so many 'religious-looking features'.[40] Nevertheless, while this feature may not be a *distinguishing* feature of religion, setting it apart from everything else, it can certainly be argued to be *characteristic* of it. Religion is a placeholder for the deeper human quest for meaning, made auspiciously vibrant through the delight of a dawning realisation that there is indeed an object of our yearning – something beyond us which somehow corresponds to our deepest intuitions and feelings.

Yet this recognition of the capacity of religion to create and sustain systems of meaning naturally raises a question. Is there some way in which science and faith can weave their narratives into something greater, with an enhanced or enriched capacity to make sense of our world and our lives?

Weaving a Richer Vision of Reality

The leading sociobiologist Edward O. Wilson (born 1929) has long argued for the need for *consilience* – the ability to weave together multiple threads of knowledge in a synthesis which is able to disclose a more satisfying and empowering view of reality. 'We are drowning in information, while starving for wisdom. The world henceforth will be run by synthesizers, people able to put together the right information at the right time, think critically about it, and make important choices wisely.'[41]

If Wilson is right, and we are indeed 'starving for wisdom', how can we become wise about the great questions of life, rather than merely informed about how our universe seems to function? We seem to be like people who know how a piano works, but can't actually use it to play a melody. As Wilson rightly observes, we need to *synthesise* – to weave together insights, uncovering a deeper and richer vision of humanity which can guide and inform our life in the present, and our hopes for the future.[42] To do this, we need to build on the core notion of a 'narrative of enrichment', such as that which I set out and defend in detail in my earlier work *Inventing the Universe* (2015).[43]

That's our agenda in this book. It is steeped in the rich traditions of enquiry and reflection we find in both the natural sciences and Christian theology, while encouraging expansion of this vision of reality through every appropriate means. It aims to open up some of the deepest and most pressing issues about human identity, welcoming scientific insights on the one hand, while aiming to develop a 'big picture' of human nature which transcends the limits of the natural sciences on the other. It does not deny any of the valid outcomes of scientific research, except the simplistic idea that these offer us a complete account of reality.

So let's begin by thinking a bit more about this strange world in which we find ourselves. Perhaps its strangest and most puzzling occupant is the human being. Let's begin to reflect on the puzzle of human identity.

2

Who are We? Wondering about Human Nature

'To know myself beloved, to feel myself beloved on the earth.'[1]
Raymond Carver

Human beings are strange creatures. We want to matter. To feel that we are loved. To be fulfilled. To be special. To achieve our potential. We pursue these goals in all kinds of ways, not really understanding why we feel driven to pursue them, yet knowing that achieving them might bring fulfilment and meaning to our lives.[2] These longings, aspirations and interpretations of life arise within us in ways that are not forced or contrived, but somehow seem to us to be inevitable and proper. In short: they are in the first place *human* and in the second *natural*.

Why Get Preoccupied With Ourselves?

Some would – perhaps not without good reason – express concern with this human preoccupation with our own situation and significance. Isn't this really a form of narcissism, which needs to be challenged and corrected? Surely we should be looking outwards at the world and its many problems that need sorting out, rather than indulging in this kind of self-important and self-preoccupied navel-gazing? It's a fair point. But what if critical reflection led us to grasp and embrace some hard truths about ourselves, which force us to give up on any delusions of grandeur and face up to ourselves as we really are?

That was certainly the view of Sigmund Freud, who declared that human narcissism had been deflated and discredited in the modern age.[3] Three geniuses had stripped away our delusions of grandeur and finality, inflicting fatal wounds on our inflated sense of self-importance. We used to think that the world revolved around us. Then Copernicus forced us to realise that the earth does not stand at the centre of the universe. We are on the periphery of things, the inhabitants of an insignificant planet.

Yet more was to come. We took comfort in the idea that, while human beings weren't actually the central focus of the universe, we were at least special on planet earth. We were the supreme rulers of the planet, utterly distinct from every other living species. Then Darwin came along, and challenged us to realise that we are part of the animal kingdom. Humanity does not even have a unique place on the planet earth.

Both these points had been made by earlier writers.[4] Freud, however, went further and declared that a third genius had come along, and wounded our self-esteem still further. Humanity is not even the master of its own limited realm, but is the prisoner of hidden unconscious forces, subtly influencing our thinking and behaviour. And who was this third genius? None other than Sigmund Freud himself.

According to Freud, each of these revolutions added to the pain and wounds inflicted by its predecessor, forcing a radical re-evaluation of the place and significance of humanity in the universe, deflating human pretensions to grandeur and uniqueness. Yet while Freud's analysis may indeed challenge our assumptions, it does not answer our questions. For many scholars, he has overstated the impact of these scientific developments. We now know that the recognition by Copernicus and Kepler that the earth is a planet orbiting the sun does not entail any diminution of the status of humanity. Some writers of the eighteenth century – such as the German poet Goethe – may have thought so. But not Kepler himself, who ingeniously (though not entirely persuasively) argued that the recognition that the earth is a planet subverted any suggestion that the earth and its inhabitants were 'below' or 'inferior to' the sun or planets.[5]

But whatever we make of Freud's assessment of our self-understanding, and his own role in this process, there can be little doubt that we need to be honest about ourselves. It's far too easy for us to take refuge in consoling stories about ourselves which depict us as supremely rational and moral beings who stand at the centre of all things. As the philosopher Iris Murdoch emphasised, we are 'anxiety-ridden animals' who try to deny the unbearable truth about our failings and delusions, and spin stories of meaning that isolate us from disturbing insights about our motives and failings.[6] That's the point that the French writer Albert Camus was getting at when he suggested that we humans are creatures who spend our lives trying to convince ourselves that our existence is not absurd.

So where shall we start our reflections? Perhaps we might begin by thinking about a model of human nature which has secured wide acceptance in some parts of Western culture.

Nothing but Atoms and Molecules

'We're nothing more than atoms and molecules! Get used to it.' It was the punchline of a rather dull lecture I once attended in London. Fortunately, I cannot recall either its title or the speaker, although I can still recall my despair over the poverty of the lecture's content and its delivery. It was, we were promised, to be a scintillating and compelling presentation of the latest scientific insights on the riddles and enigmas of human beings, which would sort out all the great questions of life. In the end, it was little more than a dull and derivative monologue, rather like warmed-up leftovers from a meal that someone else had prepared. As became clear from the speaker's glib and superficial responses to the questions afterwards, the big questions of life were not sorted out that evening, but remained tantalisingly open. They simply could not be reduced to the level of 'atoms and molecules'.

Those who resist this kind of inflated reductionism are often declared to be unscientific fools, trapped in a religious mindset opposed to the deliverances of scientific orthodoxy. Yet many of those who protest against this trend are atheist scientists who are appalled that such a misrepresentation should have become scientific orthodoxy in the first place. Raymond Tallis is a good example of a leading atheist who regards such views as indefensible and dehumanising. 'I am an atheist humanist; but this does not oblige me to deny what is staring me in the face – namely, that we are different from other animals, and that we are not just pieces of matter.'[7] Others would support this, pointing to various aspects of human culture – such as religion – as marking us off from others.[8]

Yet the closing words of that London lecture set out, clearly and precisely, the great unexamined orthodoxy of our day: that a purely scientific account of human nature and identity is possible, which makes philosophy, religion and the humanities irrelevant and outdated.[9] In its most aggressive form, this dogmatic mindset asserts that a scientific explanation is the only valid explanation and account of anything – including the deepest questions about human nature. Its main target is not, as is usually asserted, *religion*, but *philosophy*. Edward O. Wilson dismissed philosophy as consisting mostly of 'failed models of the brain'.[10] Stephen Hawking declared that 'philosophy is dead', allowing scientists to fill this gap and become 'the bearers of the torch of discovery in our quest for knowledge'.[11]

Now a fair point is being made here, even if it is overstated. To give one example: modern psychology has discredited the naïve notions of human rationality that gained credence in the 'Age of Reason' – for example, by showing how much that we like to think is 'rational' is really intuitional. Yet that's good for philosophy, because it forces it to think harder about how the human mind works. However, while a psychologically chastened philosophy would be wise to avoid *a priori* reasoning or conceptual analysis, it can be genuinely helpful in offering empirically informed reflections on critical issues.[12] We need philosophy as a critical tool; it has multiple failings when it believes it can provide indubitable answers to the great questions of life – such as the meaning of our existence.

So in what way are human beings different from other animals? What conceptual toolkits can we use to explore this question? Let's look at one, which has the potential to open up some important ways of understanding human nature, and helps us to avoid the massive shortcomings of reductionist accounts of the human.

Why We Need Multiple Perspectives on Human Nature

There are multiple aspects to human nature. Many of my academic colleagues at Oxford are involved in medical research. Each focuses on a different part or aspect of human beings. It's sometimes difficult to appreciate that psychologists, cardiologists and oncologists are studying the same human beings. They each focus entirely on one specific aspect of human wellbeing, which is their specialist area of knowledge and research. Yet there is more to human nature than any one of these areas. Somehow, we have to work out a way of recognising the complexity of humanity, without losing sight of the fundamental unity of a given human being.

That's why many philosophers and scientists are drawn to the idea of recognising human nature as a complex reality with many aspects or perspectives, realising that a failure to respect complexity leads to over-simplification and distortion. Any single perspective or viewpoint is likely to be partial and limited. It is only by recognising and integrating multiple perspectives on humanity that we can hope to understand ourselves as a whole.

A good example of this kind of approach is seen in the writings of Charles A. Coulson, Oxford University's first Professor of Theoretical Chemistry. Coulson knew the importance of good analogies in helping people to grasp difficult ideas, and developed several such analogies in affirming the complementarity of multiple perspectives on reality afforded by science and the humanities – such as poetry and religion. One of the best of these seems to have arisen out of his love of mountain walking, which developed during his period as a lecturer in Scotland during the Second World War.

Coulson realised how the complex topography of the Scottish mountain Ben Nevis could serve as an analogy for the need for multiple perspectives on life. Assuming that many of his readers would be familiar with Ben Nevis, Coulson invited them to join him in an imaginative walk around the mountain, and reflect on what they saw. Seen from the south, the mountain presents itself as a 'huge grassy slope'; from the north, as 'rugged rock buttresses'. Those who know the mountain are familiar with these different perspectives. It's the same mountain, yet a full description requires these different perspectives to be brought together, and integrated into a single coherent picture.[13]

Coulson's core insight is that 'different viewpoints yield different descriptions'. The scientist might thus stand at the north side of the mountain, the poet at the south, and so on. Each of these observers reports on what they experience using their own distinct language and imagery.[14] 'Each looks at the mountain; each sees certain things and each tries to describe his encounter with the mountain in terms that make sense. Each devises a language that is suitable for his particular purpose.' So where one observer might see grassy slopes, another might see a rocky mountain. Yet both are representative and legitimate viewpoints of a complicated natural feature.

For Coulson, this makes the need for an overall, cumulative and integrated picture of reality essential. 'Different views of the same reality will appear different, yet both be valid.'[15] The analogy is easily applied to the relation of science and the humanities, and to the various disciplines which engage with human nature. It is only someone 'who cannot, or will not, look at it from more than one viewpoint who claims an exclusive authority for his own description'.[16] Each provides only a partial account of a greater reality.

So how does this model help us think about the complexity of our existence? Let's apply it to various ways of understanding human nature, and see how it works out.

Resisting Reductionism; Affirming Complexity

I was dining with a scientist colleague at his Oxford College. It was a rather splendid meal, and I remarked on this. My colleague laughed. 'Maybe we're all just very sophisticated metabolic processing machines!' He didn't mean me to take the remark seriously. Yet it opened up a line of thought for me as I walked home that night. Yes, we are metabolic processing machines. If we were unable to convert the proteins and carbohydrates of our everyday food into energy and the basic building blocks of our bodies, we couldn't survive. Metabolism is essential to life.

Yet that doesn't mean that we are *only* metabolic machines, as if that provided a total description of a human being. It simply (and rightly) recognises that *one aspect* of our identity is our capacity to process food. Yet this human capacity to convert food into energy sustains a series of more complex and meaningful operations, such as the quest for meaning, and showing love to others. Metabolism is not an end in itself. It is the means by which some of the most significant characteristic features of human beings can be resourced. It is the means to these ends, not an end in itself. Metabolism is going to be an integral part of any full account of human nature. Yet it is emphatically not the *whole* picture. Being able to metabolise allows human beings to do more interesting things – and it is *those* that arguably define what is distinct about us.

The same issue can be seen in another familiar model of human nature. This highly reductive model is found in the writings of the biologist Francis Crick, who defined human beings in purely neurological terms. ' "You," your joys and your sorrows, your memories and your ambitions, your sense of personal identity and free will, are in fact no more than the behaviour of a vast assembly of nerve cells and their associated molecules . . . You're nothing but a pack of neurons.'[17] It's a bold statement that some clearly find enthralling. Its simplicity seems to pull out the rug from under all kinds of philosophical and theological debates which they find irritating. Human beings can be defined simply and neatly in terms of one of our physical components, which plays such a critically important role in our lives that we can treat it as determinative and identity-giving.

But we need to appreciate that Crick's highly reductive approach to our human identity assumes that a complex system is no more than the sum of its parts – and that one of these components can be singled out as being of defining importance. The most generous way of understanding Crick's hopeless overstatement is to suggest that it is a neurologist's perspective on human nature, which somehow manages to ignore the obvious fact that there is a lot more to human nature than neurons. Of course we need these if we are to function properly. But we are not defined exclusively, or characteristically, in this way. There's more that needs to be said.

Richard Dawkins' 'gene's eye' view of human nature attracted a lot of attention back in the 1980s, although it has since fallen out of favour.[18] This approach sees human beings essentially as machines which are controlled and determined by our DNA – the complex biological molecule which transmits genetic information. 'DNA neither cares nor knows. DNA just is. And we dance to its music.'[19] Our sole purpose in existing is to pass on our genes to future generations. Human beings are just gene-perpetuating machines.[20]

Once more, the problem is that of privileging one perspective of reality, and over-stating its importance. Dawkins is right to argue that human beings want to pass on their genes (even if they might not use this specific way of talking about this urge or instinct). But it's just one aspect of our complicated identity. Again, it is part of the picture – but it cannot conceivably be seen as the *whole* picture.

Another reductionist myth that has gained some popular traction is the idea that human love is just the behavioural outcome of our hormones. Now there is an obvious truth in this. There is clearly a correlation between vasopressin receptors and pair bonding in males, and probably between oxytocin and pair bonding in females. Love is indeed a drug – a heady chemical cocktail of hormones which drive us to seek out mates.[21] But that's not the full story.

The Russian writer Alexander Men (1935–90) helps us to see just how much more needs to be said here. Drawing on a story by the nineteenth-century Russian dramatist Nikolai Gogol, Men asks us to imagine Afanasy Ivanovitch, a withered old man in his eighties, who lost his wife years ago. He still bursts into tears when he remembers her.[22] He may have lost his hormonal drive, yet something deeper remains present and active. As C.S. Lewis pointed out in his late writing *The Four Loves*, erotic love is only one aspect of human relationality; it is supplemented with other forms of love – such as 'affection' – which shape our behaviour, and help determine what a truly human life might look like.

Anthropology: The Scientific Study of Human Nature

Some will rightly suggest that we should turn to anthropology – the science of human nature – to help us understand our true identity. After all, the natural sciences are one of the most reliable sources of human knowledge. Anthropology ought therefore to be able to unlock the secrets of human identity and significance with ease and precision, in much the same way that biochemistry is able to make sense of the human digestive processes. Since there is a dedicated science of human nature, we should surely take its secure findings with the greatest seriousness.[23]

Yet things turn out to be rather more complicated than might seem to be the case. For a start, like any discipline, anthropology goes through phases. Ideas that dominate the discipline in one age then fall out of favour, to be replaced with new ideas and approaches. It is fatally easy for a non-anthropologist to fail to appreciate this point, and base her ideas about culture or religion on an outdated and discredited anthropology. One of the greatest and most puzzling weaknesses of Richard Dawkins' *God Delusion*

(2006) was his baffling decision to accept as normative the outdated anthropological account of religion in James Frazer's *Golden Bough* (1890). While Frazer's views were influential in the period before the First World War, they are now regarded as discredited.[24] It remains unclear why Dawkins chose to make his own views on religion so dependent on this outdated and discredited account.

There are important debates within the discipline of anthropology itself over its nature and scope, and whether it indeed can be considered to be an 'empirical' discipline in the first place. Does anthropology study material cultures using an implicit behaviourist paradigm from outside; or does it describe symbolic cultures from within? What is the role of theory in observation? The recent critical assault on the presuppositions, methods and conclusions of Margaret Mead's ethnographical bestseller *Coming of Age in Samoa* (1928) – which defined for many what anthropology was all about – does not discredit the discipline, but does raise important questions about the status of external observers and their agendas.[25]

Anthropology cannot avoid interpretative elements – as when an external observer of a culture tries to make sense of what is observed, yet in doing so interprets that culture using presuppositions derived from their own cultural context.[26] This is not about the 'scientific' analysis of culture, but the interpretation and evaluation of one culture from the standpoint of another culture (typically a secular Western mindset).

Anthropology is superb at identifying common features of human culture – such as the virtually universal human tendency to use stories as a way of organising memories of the past, and sustaining the identity of individuals and communities.[27] But it can't make normative judgements (e.g., which of these stories is 'right', or whether this basic human tendency is itself fallacious and misguided). In this work, I shall draw on anthropological studies, as appropriate, as a description of human nature and culture, particularly in noting certain universal tendencies – such as the phenomenon of religion, and the use of stories to remember the past, and make sense of the present.

The Need for a Bigger Picture of Human Nature

Reductive views of humanity represent a single aspect of human existence as if it were the totality of that existence – or at least the aspect that really matters. Often, such approaches treat one element of the human body which supports life as if it were the ultimate reason for life itself. Yet this view surely needs to be challenged. For a start, it overlooks the relational and social aspects of human life. Human beings need to exist in relationships. As Aristotle pointed out more than two thousand years ago, human beings are social animals. It's a vitally important aspect of human existence. Yet it is only part of a more complex picture, not something which can be isolated from the remainder of human life, and treated as if it were the essence of human nature. The best way of challenging these inadequate and partial approaches is not to get lost in their fine detail, but to set out a richer and deeper vision of human nature which includes what is good about them. Human beings are complex systems; our whole transcends our individual parts.

It is always useful to have a theoretical model that both helps us to identify the difficulty with the reductionist approaches to human nature we have just noted, and also enables us to develop a better alternative. One of the most promising approaches is the philosophy of 'critical realism' developed by the social philosopher Roy Bhaskar, which recognises the 'stratification of reality'. Human nature thus needs to be conceived as consisting of different levels or strata. Humanity can be – and, indeed, ought to be – investigated at the physical, chemical, biological and sociological levels (to mention a few of the possibilities),[28] yet none of these multiple levels is to be regarded as normative or definitive; rather, each is to be considered part – and only part – of the complex reality that we know as humanity.

This 'multiple levels' approach affirms that we are indeed made up of things like atoms and molecules, or neurons (as well, of course, as many other constituent elements). This does not, however, mean that we are 'nothing' but neurons, or atoms and molecules. This hopeless over-simplification just confuses a component or level within a system with the system as a whole. We must learn to consider human beings as complex totalities, which cannot be defined or described in terms of any one of their constituent parts.

To open up this discussion further, we need to look at some of the rich understandings of human nature that have emerged down the centuries, and reflect on the questions that these raise. In the next chapter, we shall consider four divergent views of human nature, and allow them to open up some of the issues that we need to consider more thoroughly in this work.

3

Human Identity: Mapping the Landscape

'People travel so that they can wonder at the heights
of the mountains, the huge waves of the sea, the broad
flow of the rivers, the vastness of the ocean, the orbits
of the stars – and yet fail to wonder at themselves.'[1]
Augustine of Hippo

'I praise you, for I am fearfully and wonderfully made' (Psalm
139:14). The complexity and beauty of the human body has long
captured the imagination of writers and artists, whether religious
or secular. Augustine saw the human mind and body as objects of
wonder, perhaps even more marvellous than the vastness of the
natural world. The great philosopher Immanuel Kant singled out
two aspects of the natural world as rich in significance. 'Two
things fill the mind with ever new and increasing admiration and
awe, the more often and steadily we reflect upon them: the starry
heavens above me and the moral law within me.'[2]

But who are we? What are we? What do we *mean*? Though
easily stated, such questions are notoriously resistant to the
clear and simple answers that some are seeking. We so easily
conform to social pressures and in-group loyalties in our reflections on human nature, preferring to overlook harsher realities.
Yet false answers do not arise merely in response to social pressures and expectations. Every answer given to the question of
human identity is necessarily false for another reason – because
it is forced to omit so much that it cannot be considered to be
either comprehensive or reliable. In affirming *something*, such
an answer will always tend to marginalise or deny *something
else*. It may be true in part, yet it is false as a whole because of
what it excludes.

Now most of us would agree that the falsity of incompleteness is greatly to be preferred to the falsity of duplicity. Nevertheless, any pretension to offer a complete account of humanity has the potential to mislead, in that it invites collusion with a discredited reductionism. While simplicity is doubtless an intellectual virtue in many contexts, it is unquestionably a vice when it comes to human nature. One-liners on the nature of humanity might make for great newspaper headlines; they don't help us in the least to wrestle with the enigma of human identity and significance.

On Avoiding One-liner Accounts of Human Nature

Many accounts of human nature make reference to a line by Alexander Pope (1688–1744) in his *Essay on Man*: 'The proper study of mankind is man.' This is often represented as marking a decisive move away from religious to secular concerns, reflecting a growing hostility towards religious perspectives. However, even a casual reading of Pope's *Essay* makes it clear that his recommendation that humanity should study itself, rather than God, does not arise from any hostility towards religion. As our limited capacities mean that we can know so little about God, Pope argues, we ought to think about ourselves instead – because it's easier.

> Know then thyself, presume not God to scan,
> The proper study of mankind is man.[3]

Pope's *Essay* affirms the human need for faith in the face of a half-comprehended and seemingly meaningless world. The limits placed on the human intellect are such that we can only know partial truths, being able to grasp and understand only a small part of our universe. Human beings turn out to be both great and powerless, wise and foolish, and subject to significant limitations. We are 'born but to die, and reas'ning but to err', trapped in an unsettling shadowy world suspended delicately between scepticism and certainty.[4] Pope's universalisation of the category of faith thus extends it beyond the limited realm of religion to human knowledge as a whole.[5] And what of science? While lauding Newton's scientific genius, Pope deftly observes that it is one thing to describe how the universe works, and quite another to know what it – and we – might mean.[6]

So who and what we are? A number of one-line answers might be given to the question of the essence of human nature. We are rational animals; we are social animals; we are prone to violence and self-destruction; we actively seek out meaning in life. All of these individually are true, and can be shown to be true; indeed, they may even be interconnected. Some have rightly wondered whether there might be a correlation between our desire to create imaginative and spiritual identities on the one hand, and our proneness to both homicide and suicide on the other.[7]

Yet none of these elements or aspects of human nature *on its own* is adequate as a characterisation of human nature. Nor does the recognition of the accuracy of one of them entail the rejection of the others. All are part of a bigger picture; none must be allowed to define or dominate that greater understanding, which ultimately transcends and relativises its individual components. Words fail to do justice to human nature, yet we have to use words to talk about all the great issues of life, including human identity.[8] We are embodied creatures located within a limiting world, who struggle to grasp and express what lies both within and beyond that world.

And what of other aspects of being human? What of gender? Or ethnicity? In what way can the many facets of human identity, both physical and cultural, be woven together to give a coherent view of human nature as a whole, without denying the significance of individuality? Augustine of Hippo is one of many Christian writers to struggle with the 'sheer incompleteness of the endeavour of self-knowing',[9] which prevents us from securing a premature closure on some core questions of human limits.

Many of the great questions of human identity lie far beyond the scope of this work, which focuses on human beings as meaning-seeking animals. There is so much more that needs to be said!

Any quest for a greater understanding of human nature is marked by paradox and tension, rather than the slick and glib simplicities offered by colder forms of algorithmic rationalism. When Pascal famously spoke of humanity being characterised by both 'misery' and 'grandeur', he was merely pointing out the tensions experienced within human life, rather than attempting to resolve them. It proves impossible to give a scientific account of humanity that does justice to the richness of our aspirations and longings for truth, beauty and goodness on the one hand, and, on the other, our failures to live up to the standards that we intuitively believe to be right, not merely for us, but for everyone else. One of the defining enigmas of human nature is our awareness of higher values and aspirations, which we seem unable to achieve. Such transcendent goals may lie beyond our capacity, yet we nonetheless recognise them as worthy and admirable.

Given this complexity, it seems appropriate to look at some influential sketches of human distinctiveness, and reflect on how they might illuminate and inform our thinking in this study. This chapter maps the landscape of human nature by considering four influential and generative visions of humanity, each of which casts light on some aspects of our identity and challenges us to reconsider some of our easy and comfortable assumptions about our darker sides. The four figures are:

1. Augustine of Hippo (354–430), a Christian theologian now recognised as one of the intellectual giants of late intellectual antiquity, who is credited with drawing attention to the importance of the subjective experience of humanity.
2. Giovanni Pico della Mirandola (1463–94), one of the leading voices of the Italian Renaissance, whose oration on the dignity of humanity is widely seen as encapsulating this movement's vision of human nature.

3. Iris Murdoch (1919–99), a moral philosopher who believed that human moral engagement was compromised by the fundamental human incapacity to see the world or ourselves in an objective and unselfish manner.

4. Richard Dawkins (born 1941), an evolutionary biologist and critic of religion, whose *The Selfish Gene* (1976) offered an analysis of the roots of human selfishness, and suggested means by which this might be overcome.

These four dialogue partners map out some of the great questions for discussion about human nature, while at the same time disclosing the fundamental divergence and disagreement which is such a characteristic feature of human attempts to make sense of their own significance and identity down the ages.

Our four illustrative thinkers are probably better at raising fundamental questions than providing definitive answers. Yet the diversity of their emphases and responses – which reflects a comparable breadth within the broader human discussion of our nature and identity – reminds us that there are no definitive or intellectually compelling answers to these core questions. We can at best hope for justified beliefs about human nature, not proven certainties. Perhaps this absence of certainty goes some way towards explaining the amplified rhetoric and lapses into dogmatism which invariably seem to accompany any attempt to discuss religion, ethics or politics – to name only three areas of characteristically human discourse which lie beyond the realm of the provable.

Only the first of the four thinkers I have chosen to open up the question of human identity is generally considered to be explicitly religious, though two of the others recognise the importance of religion, both as a human undertaking in the first place, and as a way of framing human identity and existence on the other. In selecting these figures, I have tried to offer a broad spectrum of reflection on the issue of human nature, which opens up some questions for wider discussion, and allows other voices to make significant contributions to the wider discussion that will follow.

Augustine of Hippo

The period of late classical antiquity witnessed the slow decline and death of the Roman Empire in Western Europe and North Africa, prompting sober reflection on the fragility of human institutions and the ultimate grounds of human security. One of the most remarkable writers of this period was a Roman civil servant, who returned to his native Africa after his conversion to Christianity, and became bishop of the colonial town of Hippo Regius. It is dangerous to try to summarise Augustine of Hippo's reflections on human nature, mainly because they are so rich that there is a serious risk of reducing them to superficial platitudes. Nevertheless, the effort is worthwhile, not least because of Augustine's considerable influence in shaping both cultural and religious thinking about the mystery of human identity and significance.

Augustine's *Confessions*, written between 397 and 400, is often described as a spiritual autobiography. While this is correct as far as it goes, there is much more that needs to be said. As Augustine reflected on his own conversion to Christianity, he became increasingly aware of the importance of autobiographical memory.[10] Augustine's analysis of the nature of time emphasises both the awareness of the subjective time in which the individual thinker exists, and the awareness of the individual existing in subjective time. For Augustine, we simply cannot know ourselves completely; the fact that we are known by God thus helps anchor us in the midst of this transitory and unstable world.

Augustine developed an intellectual and imaginative framework which preserves the subjective importance of the present moment, allowing him to affirm and *integrate* the significance of his memories of the past and his hopes for the future. Recent psychological research has confirmed the importance of the notion of 'subjective time' for human identity,[11] as well as creating space for a religious or metaphysical enrichment of the notion, along the lines developed by Augustine. If Augustine has a popular successor today in this respect, it is probably C.S. Lewis, whose integration of memory and reflection is probably seen at its best in *Surprised by Joy*.[12]

Yet many would argue that Augustine's most important reflections on human nature concern sin – a notion, sidelined by many modern self-congratulatory visions of humanism, which tries to articulate what is *wrong* with us, and what can be done about it. Augustine is easily depicted as a moral and social pessimist, overwhelmed by the social turmoil of his age. Yet this is unfair. Augustine's concern is to expose the consequences of human beings existing within an historical process which limits our capacities and ability to think and act. We cannot see things as they really are. Our eyes are dimmed and unfocused, requiring both divine healing and training if they are to see our world and ourselves properly.[13] And even if we could achieve such a proper vision, we are trapped within our own self-referential schemes of thought, that dispose us not to want to accept, still less act upon, such a true knowledge.[14]

We need a fundamental realignment and recalibration if we are to live rightly and wisely – and this is something that we cannot do unaided. Augustine's emphasis on God as the one who heals our souls and enlightens our minds captures something of the human problem, while at the same time expressing the difference that God makes to human existence and experience.

Why is this so important? Like a prophet speaking from beyond the grave, Augustine challenges the naïve and optimistic view of human nature that continues to entrance some within Western culture. Some believe that we can see things as they really are; others that we are masters of our own souls and destinies; others that we are fundamentally good, and are unproblematically capable of doing what is right once we know what is right. Augustine would see all of these as understandable delusions that fail to come to terms with our own human natures. We shall explore these themes further later in this work.

Giovanni Pico della Mirandola

The European Renaissance is regularly and rightly highlighted by cultural historians as a remarkable period of creativity, reconstruction and renewal. What many regarded as a stale and decaying culture was brought to new life through a deliberate and imaginative process of cross-fertilisation, in which the artistic and aesthetic glories of the ancient world were transfused into the veins of Western European thought. To its critics, this process of reappropriation was highly selective, even arbitrary; to its supporters, it was in effect a filtering process that allowed the best of the wisdom of the ancient world to enrich a stagnant modern age.

Any attempt to single out one specific work of this age as somehow encapsulating its essence is fraught with danger, not least on account of the obvious difficulty of identifying precisely what the essential core of the Renaissance might be.[15] Yet many would still point to Giovanni Pico della Mirandola's *Oration on the Dignity of Humanity* (1486) as a brilliant vignette of the core vision of the Renaissance.[16] Although only the opening section of this oration deals with human nature and identity, the ideas it develops resonated with many at the time, as they do today.[17]

Pico delivered this flamboyant oration at the age of twenty-four. Often dubbed the 'Manifesto of the Renaissance', it was written in a highly polished and elegant Latin style, affirming a traditional Christian doctrine of the creation of humanity, yet giving a particular emphasis to one of its core elements – the creativity of humanity. Humanity's position within the created order is not fixed, but is determined by what individual human beings choose to enact.

Pico argues that God created the 'Great Chain of Being', in which every creature was allocated a specific place. Following Augustine of Hippo, Pico suggests that God, having created such a beautiful and complex universe, longed for someone who might take delight in its beauty, and be impressed by its grandeur of scale. God thus created humanity. Yet every place in the great 'Chain of Being' – from angels to worms – was already allocated. There was no missing link, no gap into which humanity might be inserted. God therefore made the decision to allow human beings to *determine their own place within the created order*.

Humanity was thus created by God as a 'creature of indeterminate image', with the capacity and permission to determine its own place in the greater order of things. For Pico, humanity has been endowed with the active capacity to determine its own identity, rather than being obliged to receive this passively in any given or predetermined fixed form. The oration emphasises the pre-eminence and unique potentialities of human beings, which mark out the human species as unique on earth.

Pico expresses this idea in a somewhat free rewriting of the second creation account of the book of Genesis, in which he portrays God as addressing Adam in the following way:

> We have given you, Adam, neither a fixed dwelling place, nor a form that is yours alone, nor any function that is peculiar to you alone. This is so that you may have and possess whatever dwelling place, form, and functions that you yourself may desire, according to your longing and judgment. The nature of all other beings is limited and constrained within the bounds of laws prescribed by us. You are constrained by no limits, and shall determine the limits of your nature for yourself, in accordance with your own free will, in whose hand we have placed you.[18]

It is thus the God-given privilege and responsibility of humanity to determine its own place and function within the 'Great Chain of Being', through the proper exercise of its freedom and intelligence. Human beings can thus choose to act as an animal, by following their lower instincts; alternatively, they can function as angels, by acting according to their higher instincts.

Why is Pico's approach so interesting? The standout point of his vision of humanity is that we have the capacity to determine what we are (or what we might become). This echoes a theme familiar to all scholars of classical Greek culture – the famous maxim in Pindar's Second Pythian Ode, written about 475 BC: 'Become what you are.' Many now follow the philosopher Martin Heidegger's influential misreading of this phrase: 'Become what you happen to be, not what you think you ought to be.' While Heidegger's reading of Pindar chimes in with postmodern values, Pindar really meant something like this: 'Become such as you are, having learned what that is.'[19] For the great thinkers of the classical age, we each have a *telos* – a goal that we are meant to achieve, which defines our true self. The 'good life' is lived out in trying to discover what that self is, and thus to become who we truly are.[20] We have to choose between different visions of ourselves, recognising that this choice is of decisive importance in shaping how we act and think.

Pico's optimistic vision of the human capacity to choose our own vision of our identity has a certain romantic attraction. But if he is right, there might turn out to be a dark side to our humanity. Let's look at just one problem that arises. What happens if someone else chooses that identity for us, imposing upon us their decisions? G.K. Chesterton put his finger on the problem here: 'When once one begins to think of man as a shifting and alterable thing, it is always easy for the strong and crafty to twist him into new shapes for all kinds of unnatural purposes.'[21] If human nature is malleable, the powerful will merely reshape it to serve their own interests – as in the 'hatcheries and conditioning centres' of Aldous Huxley's *Brave New World* (1932).

Iris Murdoch

Earlier, we noted the quote from the ancient Greek philosopher Pindar: 'Become what you are.' Another maxim of the ancient world was inscribed in the Temple of Apollo at Delphi: 'Know yourself'. This piece of advice seems to have been a commonplace in Greek literature of the classical age, and was often understood (for example, by Plato) as a criticism of speculative knowledge about the world or the gods. If you didn't even understand yourself, how could you hope to understand these greater matters? The beginning of wisdom lay in knowing ourselves, and acting on that knowledge.

It is, of course, an important piece of advice. Yet it all too easily becomes an empty platitude. What if we refuse to accept such a knowledge of ourselves, preferring an illusion to the reality? What if we prefer to construct our own identities, like the autobiographer who creatively and generously rewrites her personal history to portray herself as someone who was constantly at the centre of everything? Or as a figure of wisdom and goodness, when the reality is actually somewhat less flattering?

The question at stake is whether humanity has both a propensity and a love for self-deception. Such a trait would be nothing new. The royal inscriptions of the Assyrian king Sennacherib (reigned 705–681 BC) about his military campaigns are clearly propaganda, rather than simple history, with an ideological agenda which modern readers might regard as amounting to self-deception within the royal household.[22] But if we have a vested interest in concealing an awkward truth, how can we learn from our failings and difficulties?

Such questions lie at the heart of the moral philosophy of Iris Murdoch, who argued that humanity is characterised by its tendency to deceive itself. We are imprisoned within a cocoon of our own making, which makes us blind to the way things really are – above all, the way we really are. Moral philosophy is called on to challenge our radical tendency towards self-deception.

Murdoch argues that the reality of the world and our own moral reality lie hidden from us, until we are compelled to see it properly. 'By opening our eyes, we do not necessarily see what confronts us . . . Our minds are continually active, fabricating an anxious, usually self-preoccupied, often falsifying *veil* which partially conceals the world.'[23]

We do not naturally see ourselves as we really are. The ability to discern the truth about ourselves is an acquired habit, something that we need to develop over time. It is 'a *task* to come to see the world as it is'.[24] Seeing things as they really are, penetrating beneath the surface of appearances, is thus both the goal and the outcome of the cultivation of perceptual attentiveness. Murdoch thus holds that 'objectivity and unselfishness are not natural to human beings',[25] and rightly notes that the Christian notion of sin offers both a vocabulary and an intellectual framework for engaging this disturbing human trait.

So how is this human self-deception to be confronted? And how can it be overcome? Murdoch's core argument is that this enslaving spell of self-preoccupation and self-deception needs to be broken, and that art and literature are the main ways through which this liberation may be achieved.[26] '[Great literature] breaks the grip of our own dull fantasy life and stirs us to the effort of true vision. Most of the time we fail to see the big wide real world at all because we are blinded by obsession, anxiety, envy, resentment, fear. We make a small personal world in which we remain enclosed.' Literature and art appeal to a 'liberated truth-seeking creative imagination', which seeks to discern and embrace 'what is true and deep'.[27] Art leads us towards 'a juster, clearer, more detailed, more refined understanding of human nature, or of the natural world which crowds upon our senses'.[28]

Art thus shows us that there is a veil placed over our gaze on the natural world and ourselves, and helps us to remove it. For Murdoch, great art shows us 'the world as we were never able so clearly to see it before',[29] allowing us a heightened perception and clearer vision of reality, and thus enabling us to engage with it more justly. 'Virtue is the attempt to pierce the veil of selfish consciousness and join the world as it really is.'[30] Murdoch's assessment of the human situation is thus that it is characterised by a habit of self-deception, which can be broken by the imaginative or moral power of great art, which allows us to see ourselves from *outside* our self-created webs of meaning, thus both exposing their falsity and creating a longing to connect up with this wider world which is not of our own invention.[31] Great art attempts to show us reality untainted by 'the intrusion of fantasy', or 'the assertion of self'.[32]

Murdoch raises important questions about human nature, especially our disturbing tendency to deceive ourselves – perhaps most obvious in the lazy and complacent contemporary cultural assumption of the fundamental goodness of humanity. There are obvious questions to be explored here. Can we really escape from a delusional view of reality by ourselves? Or do we need help? While the Christian doctrine of grace affirms the goodness and kindness of God, it also discloses the incapacity of humanity to distance or detach itself from its self-created world of deceptions. Yet perhaps the important thing here is not the solution that Murdoch advocates (although this has many insights worth noting), but her insistence that there is a problem here with human nature that cannot be overlooked, and must be engaged with.

Richard Dawkins

The British biologist Richard Dawkins published *The Selfish Gene* in 1976. The work is now regarded as one of the most influential studies of human nature published in the last generation. It is unusual, in that it popularised a scientific idea with such elegance and sophistication that it had a significant impact on wider cultural conversations of the 1980s. *The Selfish Gene* developed a series of memorable analogies and lucid explanations to explain the core themes of Darwinian orthodoxy in general, while at the same time advocating Dawkins' own 'gene's eye' view of evolution.

In the late 1960s, Dawkins came to the conclusion that the 'most imaginative way of looking at evolution, and the most inspiring way of teaching it' was to see the entire process from the perspective of the gene. This 'gene's eye' approach is best seen as a creative reworking of the 'fundamental logic of Darwinism',[33] which allows the fundamental coherence of Darwinian evolutionary theory to be grasped imaginatively. The core theme of Dawkins' approach can be summed up in a sentence from his lecture notes of 1966: 'our basic expectation on the basis of the orthodox neo-Darwinian theory of evolution is that Genes will be selfish'.[34] This idea is developed and given further substance and justification in *The Selfish Gene*: 'A predominant quality to be expected in a successful gene is ruthless selfishness. This gene selfishness will usually give rise to selfishness in individual behaviour.'[35] Human selfishness is thus an expression of an underlying genetic predisposition, over which we have no control. Even altruism, Dawkins argues, can be explained in terms of this paradigm of selfishness, in that it represents a mechanism by which genes are able to ensure their survival overall, even if some individual gene-bearing individuals have to be sacrificed along the way.

The cultural background of the age created an interest in the core ideas of *The Selfish Gene* which went far beyond their scientific significance. Dawkins' suggestion that human beings could be seen as collections of 'selfish genes' helped to make sense of some of the social and political developments of the 1980s, such as the individualist political ideologies of Ronald Reagan and Margaret Thatcher which argued that greed was good for society, as it generated an appetite for new possibilities and liberated humanity from the burdens of the past.[36] Since evolution seems to help those who help themselves, selfishness should be seen as a driving force for positive change rather than as a flaw that drags us down.[37]

Dawkins argues that our genetic history predisposes us to act in certain selfish manners. So what can be done about this? Dawkins suggests that he is like an oncologist, whose professional speciality is studying cancer, and whose professional vocation is fighting it. The future of humanity depends on resisting, not endorsing, this genetic legacy. 'Let us try to *teach* generosity and altruism, because we are born selfish. Let us understand what our own selfish genes are up to, because we may then at least have the chance to upset their designs.'[38] Our genes may 'instruct us to be selfish', but we are under no obligation to obey them.

Dawkins develops the idea that, although we are shaped and conditioned by our 'selfish genes', we can achieve a state of enlightenment in which we realise the manner in which we are trapped by our genes, and devise strategies for resisting their malign influence. Humanity is trapped in patterns of thought and action which are ultimately not of their own choosing. However, Dawkins declares, humans are able to assert their autonomy in the face of this genetic determinism. We can rebel against our selfish genes: 'We have the power to defy the selfish genes of our birth ... We have the power to turn against our creators. We, alone on earth, can rebel against the tyranny of the selfish replicators.'[39] Human beings alone have evolved to the point at which we are able to rebel against precisely the process that brought us here in the first place.

Understanding the evolutionary process thus allows us to subvert its influence, and redirect its possible outcomes. We shall explore this line of thought further in this work, especially when we consider the notion of 'transhumanism'. Can we now take charge of our own evolution, and decide what we should be?

These four authors raise important questions about human nature and identity, as well as the shape that the human future might take. Although there are interesting points of convergence, what is more striking are their points of disagreement. The outcome of human reflection on the nature and future of human beings is contested, and shows no signs of being settled. Some might see this as an indication of the futility of any attempt to reflect on these matters; others, however, will see this as an invitation to revisit and reassess these great questions. We will take the second course in this book, focusing on the human quest for meaning – to which we now turn.

PART TWO

Wondering about Life:
The Human Quest for Meaning

4

Pilgrims Seeking a 'Big Picture':
The Balcony and the Road

'We had the experience but missed the meaning,
And approach to the meaning restores the experience.'[1]
T.S. Eliot

How do we make sense of our world and our lives? Many find
refuge in the delusion that we can somehow attain a privileged
viewpoint, standing above the flux of history so that we can see
the overall course of our own lives in their proper perspective.
Most, however, realise that our view is limited because of our
insertion into the historical process. For the writers of the Middle
Ages, the human being was a *viator* – someone who was travelling
along the *via* of life, a pilgrim and sojourner in our strange world.
And as we travel, we ask questions. Some are functional. Where
can I find food and water? But some are deeper questions. Why
are we on this road in the first place? Where does it take us? What
is the point of travelling along it?

Recently I found myself reading one of the neglected classics of
biosemiotics – the study of the forms of communication and
signification found in and between living systems – in preparation
for a class I teach at Oxford University. These days, few people
read Jakob von Uexküll (1864–1944).[2] Yet his account of how
animals adapt to their environments is now seen as being of semi-
nal importance. In his essay 'A Foray Into the Worlds of Animals
and Humans', Uexküll introduced the idea of a 'web of meaning'
– an internal framework of understanding through which we and
other living creatures try to make sense of our world, and which
guides us as we journey. Just as a spider spins its threads, so we
spin our relations to the world around us, and weave these into a

45

firm web which shapes our existence.³ Uexküll called this web of meaning an *Umwelt* – a perceptual world which shapes how we experience and interpret our lives and context. Human beings are by nature meaning-seeking animals, weaving webs of meaning to guide and govern their lives.

Human beings have asked these questions of meaning since the beginning of history, and they remain as important today as ever. So how can we answer them? Some are content to accept the prevailing wisdom of their peers.⁴ In an act of tribal loyalty, they align themselves with the ideas and values of a social group which is seen to be powerful or fashionable. Truth is here seen primarily as an act of intellectual and moral loyalty to a community into which we are born, or which we choose to enter. Truth is thus collusion with a group-think – about 'what it is good for us to believe', or what we are required to believe if we are to remain in good standing with those we regard as important and influential.

It's neat and simple, the ideal creed of people whose status depends on acceptance by a social 'in-group', or of politicians anxious to retain the support of voters or party selection committees. Yet it stifles independence of thought, in that this conflicts with loyalty to a social group. So we end up locked within some cultural 'in-group', and as a result are unwilling to take seriously the views of other communities. That's why fundamentalisms of every kind (whether religious, anti-religious, political or cultural) encourage social and intellectual isolation. It's so much easier to dismiss alternative perspectives if you believe that their advocates are evil, warped or deluded. That allows you to ridicule them without taking their ideas or their personal integrity seriously, and makes it easier to argue for their exclusion from the public domain.

At a more sophisticated level, this way of thinking helped create social cohesion and conformity in the Greek city-states of the classical period, which developed their own religions and philosophies as a marker of their distinct identities. Yet, as the philosopher

Richard Rorty noted, this view began to fall out of favour in classical Greek culture, partly because it was seen as incorrigibly parochial. Who wanted to be 'confined within the horizons of the group into which one happens to be born'? Writers such as Plato championed the quest for truth without reference to the norms and beliefs of a given community. Rorty himself defended the idea that knowledge was acquired through standing within a tradition, and appropriating and honouring its ideas and values. 'What matters is our loyalty to other human beings clinging together against the dark, not our hope of getting things right.'[5] Yet surely we should try and pursue a knowledge that is independent of the culture or tradition within which we live?

We have already noted Alexander Pope's *Essay on Man* (1732–4), one of the greatest literary works of the 'Augustan Age'. This rich and complex poem is a shrewd reflection on the aspirations and limits of being human, and our difficulties in making sense of the universe within which we find ourselves placed.[6] Pope recognises that this universe appears incoherent and morally ambiguous, perhaps characterised by evil more than by good. Yet, Pope suggests, we have to take account of our frail and fallible moral and intellectual capacities in reaching this judgement. Perhaps the universe appears imperfect and incoherent to us because of the limits placed on human perception. Life seems chaotic and purposeless to us because we are immersed within the flux of things, and cannot extricate ourselves from it to catch a tantalising full glimpse of reality which alone could disclose that we have a meaningful place in a coherent universe.

This is a deeply appealing idea. It invites us to see ourselves and our world as we really are, not as we have constructed ourselves. It is about discovering the truth about our world, not inventing our own imaginary world. The philosopher Ludwig Wittgenstein realised that meaning and happiness arise when we believe that we are thinking and living in accordance with something deeper and greater than ourselves. 'In order to live happily I must be in agreement with the world. And that is what "being happy" means.'[7] We need to grasp the 'big picture' of the universe, and position ourselves within it.

Yet we seem unable to grasp this full picture by ourselves. We need help if we are to see beyond the frustrating limits imposed upon us by being human. This is, of course, a classic theme in Christian theology. The notion of divine revelation is about the disclosure of a view of reality which we did not invent, and which lies beyond the capacity of human reason to grasp fully. Revelation is not about the violation of human reason, but a demonstration of its limits, and a disclosure of what lies tantalisingly beyond its limits. It is about the illumination of the landscape of our world, so that we can see things more clearly.

For Christians, this capacity to see things as they really are – rather than as they are glimpsed from the surface – is a gracious gift of God. Our eyes need to be opened, so that what we once deemed to be an incoherence is recognised as arising out of our inability to see fully and properly. How can we affirm and hold together the 'heart-breaking brutality and the equally heart-breaking beauty of the world'?[8] While many Christians think of the Holy Spirit primarily as a source of *empowerment*, perhaps the true role of the Spirit lies rather in *discernment* – not in 'making us supernaturally strong', but in 'opening our eyes'.[9] There is indeed a 'big picture' of our mysterious world – but it is not one that we discerned on our own, still less one that we invented. It was one that had to be *shown* to us.

C.S. Lewis on Seeing Things Properly

The Oxford literary critic and apologist C.S. Lewis is one of the best-known advocates of seeing the Christian faith as an illuminating and informing 'big picture'. He expressed this idea in a signature statement, which sums up his reasons for becoming a Christian in the first place, and remaining one in the second. 'I believe in Christianity as I believe that the Sun has risen, not only because I see it, but because by it, I see everything else.'[10]

Lewis's statement needs further development to appreciate its importance. It is not an argument for God's existence, but rather the observation that we experience the rational and imaginative capacity of the Christian faith in and through the process of inhabiting it and using it to frame our observations and experiences. There is an interesting parallel here with reflections about the human mind. In a sense, we only experience our minds directly when we use them. The human mind is not something that we observe; rather, it is something that we experience through observing. It is not so much that we infer the existence of our minds; rather we infer the existence of everything else using our minds, and as a result come to appreciate the explanatory potential of this strange entity that we call 'the mind'.[11]

Although an atheist as a young man, Lewis gradually came to see this as imaginatively deficient, locked into a 'glib and shallow rationalism' which failed to do justice to the richness of the world, or the deepest longings and concerns of human beings.[12] Where some turned to religion in order to escape the complexity and uncertainty of the world, Lewis offers us a theological vision that does some justice to the intricacy of our world, allowing us to see it as orderly without being mechanical. Instead of the impersonal determinism of worldviews such as Marxism, Lewis places the personal reality of God at the heart of a universe that acknowledges our existence, and our role in redirecting its story.

For Lewis, Christianity provided a new way of seeing things, and hence a new way of experiencing the world and life. We see this in Lewis's comments on Dante's *Divine Comedy*, in which he suggested that the poet's imagery of light allows us to 'not only understand the doctrine but *see the picture*'.[13] It is about the transition from one way of seeing our universe and life itself to another. Yet this change in the way in which we see things is so great and radical that we need to think of ourselves as a 'new creation'. The old world that Lewis left behind was a world in which his status depended on his achievements and success. It was a world limited to what reason and science could disclose, devoid of intrinsic meaning and value.

Paul's New Testament letters speak of Christians having the 'mind of Christ' (e.g., 1 Corinthians 2:16), in which a 'new self is created because a new world is being entered'.[14] Lewis's emphasis is more easily accommodated to the related idea of the 'enlightenment of the eyes of our heart' (Ephesians 1:18), with its strong emphasis on the transformation of our vision of reality, as a result of the healing and renewing work of divine grace.[15] For Lewis, the way in which we act within our world is shaped by the imaginative framework through which we see it.[16]

Lewis is a strongly visual thinker, who found the imaginative worlds of narrative to be a more adequate way of exploring and explaining ideas than conceptual analysis. In 1933, Lewis published *The Pilgrim's Regress*, an allegorical account of his own conversion, and especially the difference this made to the way in which he saw the world around and within him. This early work, which many find difficult reading, is best understood as an imaginative mapping of the landscape of faith, focusing especially on the outcomes of the transition from unbelief to faith. Its central character is the pilgrim – 'John' – who has visions of an island that evokes an intense yet transitory sense of longing on his part. At times, John is overwhelmed by this sense of yearning, as he struggles to make sense of it. Where does it come *from*? What is he yearning *for*?

While Lewis's account of his discovery of faith is interesting, both in terms of illuminating his personal journey of faith and the way in which he expressed this at this early stage in his career, the most important part of the work is that which describes the 'regress'. *The Pilgrim's Regress* actually describes *two* journeys – the first as he journeys to the mysterious island, and the second as he returns home. When John walks back through the same landscape *after* coming to faith – the 'regress' of the book's title – he discovers that its appearance has changed. Though it is the same landscape, he is told that he is now 'seeing the land as it really is'.[17] John's encounter with his object of desire has transformed his capacity to see himself and his world.

The basic idea of being able to see things in different ways is, of course, not unique to Lewis, having a rich history of use within the Christian tradition (for example, in Augustine of Hippo and Dante). It was also developed by the philosopher Gustav Fechner, who had a deep influence on the composer Mahler.[18] Fechner argued that the world could be seen in two fundamentally different modes – the 'daylight view of the world', which disclosed its vibrancy and deep inner meaning, and the 'night' view, which was limited to its material aspects. The same thing is seen in different manners and to different extents, depending on the amount of illumination that is available.

We now need to think more carefully about the different perspectives we bring to bear on the world, as we try to make sense of it. In what follows, we shall draw on one of the most perceptive reflections on this issue – John Mackay's justly celebrated image of the 'Balcony and the Road', set out in 1942.[19]

The Balcony and the Road

John Alexander Mackay (1889–1993) served as President of Princeton Theological Seminary from 1937 to 1959.[20] Mackay spent some time during 1915 in the Spanish capital, Madrid. Spanish families, he recalled, would gather on the balcony of an upper room during the evening, so that they could look down on the bustling street beneath them, or at the stars above them. Mackay made this the basis for his reflections on the perspective of the human observer in achieving insight into both human nature and God.

For Mackay, the Balcony was a metaphor for the 'perfect spectator, for whom life and the universe are permanent objects of study and contemplation'.[21] Life on the street below can be observed with interest; it is certainly observed from a distance, with a sense of detachment. Science is the supreme example of such a perspective. 'The scientific laboratory is the Balcony at its best, and the scientist is the ideal spectator.'[22] Using such tools as microscopes and telescopes, the scientist aims to cultivate an 'objective view of things', from which all subjectivity is eliminated.

Mackay argues that this 'contemplative' approach to reality failed to recognise the awkward but critically important point that observers are actually part of the reality that is being observed, and are affected by the world around them.[23] Inevitably, this meant that the insights gained from the Balcony were only partial accounts of reality, and failed to do justice to its many aspects and levels.

> The scientist, however, can never achieve an interpretation of life and the universe. The deeper he probes into one sphere of reality, the slighter becomes his knowledge of the whole ... The greatest realities are such that their nature can never be known if they are treated as mere objects. Such is God, who is eternally subject, and can never be reduced to an object. Such are human beings, who are other selves that, in their inmost core, will remain impervious to the gaze of the scientist in search of objectivity.[24]

Mackay contrasted the view from the Balcony with life on the Road, characterised by 'a first-hand experience of reality'.[25] Those who are travelling on the Road are participants, not distant and detached observers. If they observe, it is in their capacity of those who are travelling alongside and amid those whom they observe. Truth is something to be *done*, not merely to be *thought*. It cannot be purely objective, in that it impacts upon life, and thus transforms it. It is existential, in that the thinker 'steps out' on the Road, committed to the journey that lies ahead.

There's another point that needs to be made here, which is not developed by Mackay himself. The Balcony is a place of privilege, in which we look at a scene that is below us. It offers, so to speak, a God's-eye view of things. The view from the Road is one of immersion. In addition to being participants in the journey, our vision is limited to what can be seen from the Road. We cannot escape from its limiting perspectives, and have to learn to work with them. The philosopher Thomas Nagel has argued that every viewpoint on the world or life is actually a 'view from somewhere'.[26] Nagel points out that we 'cannot escape the condition of seeing the world from our particular insertion in it', however much we may long for the historical and cultural detachment afforded by the Balcony. We will return to this point later, in thinking about the initial appeal and ultimate failure of some forms of Enlightenment rationalism.

We cannot rise from the road by ourselves, and attain a God's-eye perspective. But what if that perspective was somehow to be brought to us? What if someone was able to disclose to us such a big picture, despite the fact that it lies beyond the capacity of fallen and finite humanity to discern it, save in half-glimpsed and half-imagined ways? Christianity affirms that there is indeed such a greater vision of reality that lies beyond the limits of pure reason – yet which is capable of being recognised as reasonable once it is grasped. The Christian doctrine of incarnation affirms that the God who alone sees all things from the Balcony chose to come and walk on the Road with us, allowing us to glimpse and trust a bigger picture of our journey. Yet the incarnation is not simply about the disclosure of a 'big picture'; it is about affirming the presence of God on the Road as we travel.

Later in this work, we shall highlight the role of 'mental maps' of reality in human accounts of the meaning of life. So what sort of maps could be drawn from the Balcony and the Road? The 'God's-eye' view of the Balcony allows a total mapping of reality, in which every interconnection is made crystal clear. Everything can be observed and depicted. But on the Road, things are different. We can't see everything. But what we can do is tell stories about what we did, and make sketches of what we saw. Early naval navigators called these collections of wisdom a 'Rutter'.[27] The

Rutter was a distillation of personal experience from those who had crossed the seas, recording the bearings on which they sailed, the obstacles they faced, and the locations of safe harbours and sources of food and water. It was a narrative of a journey, characterised by the limited perspectives of mariners on the high seas, not a detailed representation of the landscape seen in its totality. People on the Balcony could draw a diagram of what they saw; people on the Road would share stories of what they experienced.

For Mackay, the Christian faith is about life on the Road, rather than the view from the Balcony. Faith is about active engagement in life, not passive observation from a distance. The detached objectivity of science is to be welcomed, yet it restricts the capacity of science to engage with the deepest existential questions of life, including those of meaning and value. Mackay's point is important, and easily defended. We shall explore this by looking at the idea of a 'world line', which plays an important role in Albert Einstein's theory of relativity.

World Lines and Long Pink Worms

The idea of a 'world line' is associated with the mathematician Hermann Minkowski (1864–1909), who developed an 'algebra of space and time' which identifies events by their four coordinates x, y, z and t.[28] The 'world line' is the curve connecting a series of points in space-time which represents the history of a particle or an observer. It can be seen as a classic representation of the view from the Balcony – an objective, detached account of what is happening in the universe.

But what if the world line refers to a person, rather than a particle? To a living, thinking, conscious individual, who is trying to figure out what their individual world lines might mean, rather than to a fermion or boson? This idea was explored, somewhat playfully, in 'Life-Line', the first published short story of Robert A. Heinlein (1907–88), a science fiction writer with a keen sense of the importance of deeper issues of life. Heinlein asked his readers to imagine a person's history as a 'long pink worm, continuous through the years, one end in his mother's womb, and the other at the grave'.[29]

But what lies beyond this 'long pink worm'? What lies before birth, and after death? And how does the existentially important notion of the 'present moment' – the 'Now' – fit into this? How can we distinguish the past, present and future? All can be represented *chronologically and spatially* using a world line. Yet their significance cannot be represented *existentially*. Einstein took the view that the 'distinction between past, present and future has only the meaning of a persistent illusion'.[30] Most of us, however, find it very difficult to think neutrally and dispassionately about the transition from a past in which we did not exist, through a present in which we live and think, to a future in which we will no longer exist.

For most of us, there is a critically important *subjective* distinction between past, present and future which is real, which matters profoundly to us. Our own individual world lines seem very small and insignificant when set against the backdrop of cosmic time. Yet the fact remains that, although we only occupy a tiny slice of the four dimensions of space-time, we want to live good and meaningful lives – and this means trying to figure out what things (including our own world line) actually *mean*.

The philosopher Rudolf Carnap (1891–1970) discussed this issue with Einstein during the 1940s, and came to the conclusion that Einstein believed that 'scientific descriptions cannot possibly satisfy our human needs'.[31] The human quest for meaning simply could not be met purely through physics. There were certain significant existential concerns – such as the concept of the 'Now' – which tantalisingly lay 'just outside of the realm of science'. Einstein's objective world line is the outcome of seeing our universe from the perspective of the Balcony; our quest for meaning is the outcome of seeing that universe from the perspective of the Road. We think, reflect and act within the limits and framework of Heinlein's 'long pink worm', and we cannot escape this framework.

We can, however, overcome at least some of the limitations of living within this 'long pink worm'. As Heinlein himself pointed out, these worms interconnect with others. We can transcend the limits of our own limited world lines by overhearing the thoughts and reflections of others from within their lines. Although C.S. Lewis does not frame his analysis of the importance of reading classic works of theology and literature in terms of world lines, the fundamental point he addresses is essentially the same. Lewis's essay 'On the Reading of Old Books' (1944) addresses the question of how an individual thinker can see with other eyes, without losing their individuality.[32] How can the limits of one person's specific perspective be transcended, without compromising their individuality?

Lewis's answer remains significant. It is through the reading of books that one is able to enter into someone else's world line, and to see things as they do. Through travelling along the Road, we can learn from others who have travelled before us. The wisdom they accumulate is transferable from one person to another. By stepping into someone else's reflections on their world line, we are able 'to see with other eyes, to imagine with other imaginations, to feel with other hearts, as well as our own'.[33] This process enables us to transcend our limits as subjective individuals, opening up new imaginative and reflective possibilities. 'My own eyes are not enough for me, I will see through those of others . . . In reading great literature, I become a thousand men and yet remain myself. Like the night sky in the Greek poem, I see with a myriad eyes, but it is still I who see.'[34]

Mackay's symbols of the Balcony and Road are helpful as we try to work out how to make sense of our world and our lives. In the end, we need both these perspectives and the respective tool-kits that they offer. A clinically detached objective view of the world has its merits; it is, however, existentially inadequate. A purely subjective view of the world might easily be illusory and self-serving, simply echoing our own prejudices and pre-commitments. Somehow, these must be integrated into a greater whole. Facts are not enough; we need to know what they mean if we are to secure the 'big picture' of reality which weaves together the objective and subjective, and above all which discloses the value and significance of life.

We shall explore this theme further in the next chapter.

5

Searching for Meaning:
Why We Need More Than Just Facts

'We cannot simply eat, sleep, hunt and reproduce
– we are meaning-seeking creatures.'[1]
Jeanette Winterson

We are indeed meaning-seeking creatures. Modern psychology
points to there being a universal human sense of a need for meaning.[2] Not everyone shares this experience, and not everyone who
experiences this need goes on to reflect on its implications. But for
most people, it's there, and it's important. This often takes the
form of a longing for a sense of personal coherence.[3] We want to
understand our experience and to feel that our lives have significance and purpose.[4] Psychology can't tell us what life means; after
all, that's not an empirical question. It does, however, make it
clear how enormously important this notion is for human beings,
and the difference that it makes to human flourishing and wellbeing. Meaning in life could be defined as: 'the extent to which
people comprehend, make sense of, or see significance in their
lives, accompanied by the degree to which they perceive themselves to have a purpose, mission, or overarching aim in life'.[5]

Where professional philosophers have virtually given up talking about the subject of meaning, psychologists have moved in to
map this fundamental human concern, helping us to understand
some of its core facets, and the difference it makes to life. We need
to feel that we can make a difference to things, and to take control
of our lives; we need a sense of identity and purpose if we are to
cope with traumatic experiences in life and our awareness of our
own mortality.[6] Basically, human beings actively seek for systems
of meaning which embrace an understanding of the world, our

personal significance, and our capacity to transcend our limits and locations, as we sense we are part of something bigger and greater.

Research on meaning suggests that it is helpful to draw a distinction between 'global' and 'situational' meaning.[7] *Global* meaning systems weave together beliefs, goals and subjective feelings of meaning or purpose in life to give a 'big picture' of reality. Yet we use this overarching global framework of beliefs, goals and sense of purpose to structure our lives and to assign meanings to specific experiences – in other words, to determine *situational* meaning. The 'big picture' thus frames our experiences, and helps us work out what they mean.[8]

Our theme in this chapter is the importance of meaning for human flourishing. We may not understand why, but it is natural for us to believe there is some meaning to be found, and to try and find it. At one level, we are looking for a 'global' way of thinking – a 'big picture' which interprets and illuminates our world and our experience. Yet this 'big picture' needs to be applied locally – to situations that we need to understand, to problems that we need to solve and to challenges we need to face. This 'situational' aspect of meaning is best seen in the realm of attitudes, values and behaviour, where our beliefs about the nature of things find their expression in what we actually do.

Although some 'big picture' philosophical systems such as Marxism do indeed engage with these questions,[9] it is widely agreed that religion is one of the most common and powerful sources of meaning. Religion can provide us with a comprehensive and integrated framework of meaning that helps to explain many events, experiences and situations in ways that satisfy both our cognitive and existential concerns, while providing a way of helping us to transcend our own concerns or experiences, and connect up with something greater.

While religion has many aspects, one of its most important is its capacity to provide a comprehensive framework of interpretation of experience and life that allows the discernment of meaning.[10] Religion is able to embrace and inform a vast scope of issues, including beliefs about the world (such as human nature, our social and natural environment, and the afterlife), contingencies and expectations (rewards for acting well and punishment for doing evil), goals (such as benevolence and altruism), actions and attitudes (such as compassion, charity and violence) and emotions (such as love, joy and peace).[11] These provide a sense of significance, particularly in enabling individuals to feel part of a greater scheme of things. The term 'big picture' is particularly helpful in conveying this idea of a perception of coherence within a complex world of experience, and helping individuals to relate their own situation to a greater reality.[12]

We need more than science can provide for us if we are to lead meaningful lives. Ludwig Wittgenstein was clear on this point. 'We feel that even if *all possible* scientific questions be answered, the problems of life have still not been touched at all.'[13] Science can only offer limited guidance as we reflect on the issues of meaning and value. In his critique of the growing trend in German academic circles towards scientific rationalism, the sociologist Max Weber (1864–1920) pointed out that it was difficult to see how the natural sciences could teach us anything important about the *meaning* of the world that they investigated.

So how do we deal with the complexity of our world? One of the greatest temptations is to reduce it to something neat, tidy, and above all manageable. One of the most striking weaknesses of the eighteenth-century 'Age of Reason' was its failure to respect the complexity and fuzziness of our world. Its 'obsession with reducing the muddy and mixed to the clear and distinct'[14] led to it demanding all-embracing accounts of reality which were totally comprehensive and unambiguous. Yet our multi-coloured and richly textured world of experience and observation simply could not be reduced to these categories without simplification and distortion.

A good example of something that simply cannot be reduced to 'clear and distinct ideas' is the beauty of natural landscapes.

On closer inspection, natural beauty often seems to co-exist with ugliness – for example, an attractive forest scene which seems spoiled by the presence of the rotting carcass of an animal.[15] The natural world is ambivalent, studded with beauty at some points, and suffering and decay at others. We cannot express this in 'clear and distinct ideas' without failing to do justice to the complexity of things.

For the English poet Thomas Traherne (1636–74), the human eye all too often fails to penetrate the depths of nature. It skims the surface of reality, failing to penetrate deeper, lacking commitment and discernment.[16] 'The world is a *mirror* of *infinite beauty*, yet no man sees it.'[17] To discern this beauty requires the development of a particular way of seeing things – a heightened perceptiveness and disciplined alertness, which needs to be developed through both the mind and the imagination.

There are two difficulties here. First, there are limits to the human capacity to make sense of our world. We don't have the mental capacity to grasp things fully and properly, and have to make do with the best that we can. If things seem blurred, it's because there are limits to our vision. The New Testament makes extensive use of images suggesting that fallen humanity has a defective vision, which calls out to be healed by divine grace. Paul, for example, speaks of our vision being obscured by a veil, which needs to be removed if we are to see properly (2 Corinthians 3:13–17). At times, Paul's imagery shifts, suggesting that our failure to see things clearly is a consequence of our creaturely status, not merely our sinfulness – for example, his famous remark about now seeing things dimly, as if through a mirror (1 Corinthians 13:12), in the hope that we will eventually see them fully.

The second difficulty is that some may be looking at the world through the wrong lens. A 'theory' is first and foremost a way of looking at the world. Things which actually are clear and interconnected may appear to be out of focus and incoherent because we are using an instrument of vision which is defective, and prevents us from seeing things properly. Moments of scientific breakthrough often involve a connection being established between observations that seemed to be totally disconnected. For example, Isaac Newton showed that the orbits of the planets around the sun and the path of objects falling to earth – such as his famous apple – were fundamentally interconnected, and governed by the same laws.[18]

A perception of incoherence or meaninglessness within the world can easily arise from the imaginative failure of an inadequate theory, rather than from the universe that is being observed. Richard Dawkins' statement that 'the universe we observe has precisely the properties we should expect if there is, at bottom, no design, no purpose, no evil and no good, nothing but blind pitiless indifference'[19] is simply his way of looking at things, using his 'New Atheist' instrument of vision. But what if his theoretical lens is out of focus, preventing him from discerning anything other than an inferred blind cosmic indifference?

Science and Meaning

What is the meaning of life? As we have seen, this is a very natural human question. Science can certainly help us to clarify what sorts of things people find to be *meaningful*. But that's not the same as telling us what the meaning of life is. In the end, that's just not a scientific question. It is one of those many great questions which transcend science's capacity to answer. Karl Popper, the noted philosopher of science, referred to these as 'ultimate questions' which related to the great 'riddles of existence'.[20] These questions really matter to most people. What is the point of life? What is the good life, and how do I lead it?

We've already seen how many professional philosophers tend to see these questions as meaningless or incoherent. For the professionals, these are the sort of questions asked by philosophically unsophisticated people. Now, there may be some truth in this. But there is a still greater truth here, which is that these questions *matter* to most people, even if they have difficulty in articulating them in terms that command respect within the philosophical guild.

Some do believe that science is able to answer these questions – even if the answers are somewhat bleak and challenging. The atheist philosopher Alex Rosenberg takes the view that science is 'our exclusive guide to reality'.[21] It is a difficult position to maintain, as Rosenberg himself admits, as arguments in its support tend to be 'viciously circular'.[22] To defend its approach, you have to presuppose the accuracy of its core beliefs. (The same problem arises with rationalism, which is obliged to assume the validity of reason in order to demonstrate its unique competence.) Yet for any research method to be validated properly, we need an extra-systemic vantage point from which it can be judged. However, if there is a vantage point beyond science by which it may be judged, then the exclusive authority of science is clearly called into question, in that its authority ultimately rests on something else – something that lies *beyond* science. Scientism itself cannot be judged or justified without appealing to, or presupposing, matters that transcend scientific inquiry. Curiously, scientism, rationalism and religious belief all find themselves depending on ideas whose validity cannot be absolutely demonstrated – yet which are still believed to be reasonable by those who embrace them.

Rosenberg argues that science alone is able to provide clear and compelling answers to questions that were traditionally seen as ethical, religious or philosophical. For Rosenberg, these are fundamentally *scientific* questions, and thus can only be answered by science. Rosenberg helpfully provides some examples of questions that are easily answered in this manner, along with his clear answers, to show off the elegant simplicity of his approach.

Is there a God? *No.*
What is the purpose of the universe? *There is none.*
What is the meaning of life? *Ditto.*
What is the difference between right and wrong, good and
bad? *There is no moral difference between them.*

Science here debunks the pretensions and mystifications of theology and philosophy. Yet on closer examination, this approach is not about science, but about the application of a specific philosophical interpretation of science which goes way beyond anything that the scientific evidence demands. It smuggles in all kinds of metaphysical assumptions that may seem self-evident to those who think like this, but which are not demanded or warranted by the evidence. We see here not a loyalty to science or evidence, but to conclusions that are prior to that evidence, and used to control and constrain that evidence. Many scientists will squirm with intellectual discomfort and irritation at this obvious philosophical hijacking of science.

Many, however, will feel that the conclusions reached call into question the methods used to derive them. Is there really 'no moral difference' between right and wrong? Between good and evil? These simplistic statements seem shallow and empty as we contemplate 'horrendous evils' – such as Nazi extermination camps, torture chambers, and the mass graves of women and children massacred for belonging to the wrong tribe. Rosenberg's approach deprives us of a vital element of the moral framework that we need to challenge evil – to *name* it for what it really is.

If Rosenberg is right, it is meaningless to speak of an action or attitude as being 'evil'. For Rosenberg, science cannot tell us what is good or evil. This view is widely held among moral philosophers, who rightly note that there cannot be a deductively valid argument whose premises take the form of factual or scientific statements, and whose conclusion contains a moral statement.[23] Rosenberg goes further, and suggests that, for such reasons, good and evil are essentially meaningless terms. So if we end up inventing our ideas of good and evil, why not do the same with our ideas of meaning? We'll come back to this important question later (73–90), as we consider whether the 'meaning of life' is something discerned or invented.

Why Facts Aren't Enough

'In this life, we want nothing but Facts!' It's a great strapline. But what sort of life can be based solely on mastering facts? One of the more memorable characters created by Charles Dickens was Mr Gradgrind, a dour schoolteacher in *Hard Times* (1854). His educational philosophy was simple. 'Now, what I want is, Facts. Teach these boys and girls nothing but Facts. Facts alone are wanted in life. Plant nothing else, and root out everything else. You can only form the minds of reasoning animals upon Facts: nothing else will ever be of any service to them.'[24] For Gradgrind, human emotion and imagination were useless distractions to the business of number crunching. Facts alone gave rise to intellectual certainty. Yet if we were mad enough to follow Gradgrind's philosophy, we would simply end up being deluged with information, while failing to find wisdom.

Facts just aren't enough. They are the raw material of knowledge, but they need to be interpreted and understood. Even science is about more that the collection of factual observations. The great Renaissance philosopher of science Francis Bacon made this point back in 1620. Some thinkers, he suggested, were like ants – they just accumulated things. Others, however, were like bees, who gathered material from the 'flowers of the garden and of the field', but transformed it into honey. Bacon's point was simple: science is about understanding the significance of facts, and trying to discern the bigger picture which lies behind them. It is about the development of *theories* – ways of seeing reality which make sense of our factual observations, while going beyond them.

A good example of this concerns the movement of the planets against the fixed stars. What is the best explanation of these movements? The traditional explanation was that the sun, moon and all the planets revolved around the earth. This 'big picture' was not itself an observation, but was rather an *interpretation* of observations. Yet as these observations became more accurate, doubts began to arise about its reliability. Might these observations of planetary movements be accounted for more effectively if it was proposed that the earth, along with other planets, revolved around the sun? By the middle of the seventeenth century, this heliocentric way of picturing things had gained the ascendancy.

Each scientific observation of the position of a planet against the fixed stars was a fact – important in itself, yet failing to extend the reach of our knowledge. The key question was this: was there a 'big picture' which accounted for these facts, yet extended our knowledge by going beyond a mere accumulation of information? If each observation was a thread, could these be woven into a fabric of meaning? Mr Gradgrind was only concerned with facts; he does not seem to have grasped that what we really need lies beyond them – even if it includes them. Gradgrind's joyless rational world consists of cold facts which marginalise – if not totally exclude – the imagination and emotions.

Dickens deftly brings out the utter lifelessness of this factual world through the lens of Gradgrind's daughter, Louisa. She is portrayed as having a 'starved imagination',[25] inclined to view everything from the standpoint of 'reason and calculation'.[26] Yet her impressive accumulation of facts and figures does not equip her to live out a joyful life. She needs something deeper to bring stability and joy to her life. Having been indoctrinated into her father's creed of the all-sufficiency of cold, hard facts, Louisa finds herself trapped in a loveless world, unable to find happiness and security.

So let's be clear: facts matter. They are the building blocks of the 'big pictures' which give life meaning, purpose and joy. Yet what really matters in life is not a wearisome accumulation of facts, but the discovery of the greater reality that lies behind them. Ludwig Wittgenstein realised this in 1916, when he penned these words: 'To believe in a God means to understand the question about the meaning of life. To believe in a God means to see that the facts of the world are not the end of the matter.'[27]

Human beings need more than facts; they seek meaning. And very often they express this meaning using stories.

Telling Stories of Meaning

One of the most effective ways in which human beings preserve and transmit insights about meaning is by telling stories. These appeal to our imaginations, while at the same time pass on information about the past and its importance for our present. Telling stories allows narration to be supplemented with interpretation. This is what *happened* – and this is what it *means*. Not surprisingly, this important aspect of human life has been noticed and explored by cultural anthropologists (24–25), who have highlighted the importance of this apparently universal human tendency to tell stories. '[We] are animals who must fundamentally understand what reality is, who we are, and how we ought to live by locating ourselves within the larger narratives and metanarratives that we hear and tell, and that constitute what is for us real and significant.'[28]

We tell stories to make sense of our individual and corporate experience – whether this 'sense-making' is stated in political, religious or more general terms – and to transmit these ideas within a culture.[29] Stories provide a natural way of organising, recalling and interpreting experience, allowing the wisdom of the past to be passed on to the future, and helping communities to gain a subjective sense of social or religious identity and historical location.[30] They can consolidate the identity of a community, just as they can help rebel poets and artists create their own micronarrative of the world through rejecting the metanarratives of their age.

But *why* do we tell stories in this way? Why are we storytelling and meaning-seeking animals?[31] If storytelling is a fundamental human instinct, what story can be told to explain this? Is there a framework of interpretation that helps us to make sense of it? Carl Jung famously suggested that there were certain 'universal psychic structures' which underlie human experience and behaviour – an idea taken up in Joseph Henderson's famous account of the fundamental plotlines of stories, such as the 'myth of the hero'.[32] Tolkien, however, saw the human instinct to tell stories of meaning as grounded in a Christian doctrine of creation, and offered a theological explanation for our love of narration.

Tolkien argues that our capacity to create stories such as the great fantasy epic of *The Lord of the Rings* is the result of being created in the 'image of God'.[33] 'Fantasy remains a human right: we make in our measure and in our derivative mode, because we are made: and not only made, but made in the image and likeness of a Maker.'[34] Tolkien is often described as developing a 'theology of sub-creation', in that he holds that human beings create stories which are ultimately patterned on the 'Grand Story' of God. We unconsciously tell stories which are patterned along the lines of this great story of creation and redemption, and which reflect our true destiny as lying with God. For Tolkien, one of the great strengths of the Christian narrative was its ability to explain why human beings tell stories of meaning in the first place.

Story, Meaning and Literature

It is only to be expected that many in the world of literature both celebrate the importance of stories and reflect on their deeper significance. J.R.R. Tolkien's literary work in the field of Nordic culture and literature persuaded him of the importance of what he termed 'myths'. While most people understand 'myth' as 'a false story' or an 'outdated account of something', Tolkien used the word in a more specialised sense. For him, it referred to a narrated account of reality, which appeals primarily to the imagination and secondarily to the reason. A myth is a story that conveys meaning, capturing our imaginations and informing our reason. Tolkien himself put this idea to good use in his celebrated *Lord of the Rings* trilogy.

Tolkien knew there was another respect in which the Christian narrative helped to make sense of the world: its capacity to explore the relation of other faiths and worldviews to Christianity. Tolkien himself was particularly interested in how the great Nordic pagan myths related to the gospel. He argued that the Christian 'Grand Story' was able to show how pagan myths arose in response to a deeper truth, yet represented inadequate and imperfect realisations of that truth. Human 'myths' allow a glimpse of a fragment of that truth, not its totality. They are like splintered fragments of the true light. Yet when the full and true story is told, it is able to bring to fulfilment all that was right and wise in those fragmentary visions of things.

For Tolkien, the Christian 'myth' created intellectual and imaginative space for other stories. By offering what Tolkien termed 'a far-off gleam or echo of *evangelium* in the real world', pagan myths could create both an appetite and an opening for the discovery of the deeper truth that underlies all truth, however fragmentary and veiled.[35] Christianity, rather than being one myth alongside many others, is thus to be seen as representing the fulfilment of all myths – the 'true myth' towards which all other myths merely point; Christianity tells a true story about humanity, which makes sense of all the stories that humanity tells about itself. Some readers will not be persuaded by Tolkien's ideas, feeling that they are perhaps too optimistic about what can be glimpsed of God from the human side of things. Yet for Tolkien, a love of pagan myth could become a gateway for the discovery of Christianity.

The religious importance of such narratives can be seen from what is perhaps one of the best-known biblical stories, which tells how the people of Israel were delivered from their harsh bondage under Pharaoh in Egypt, and led through the wilderness of Sinai into the 'Promised Land' of Canaan. The Passover celebration recalled the events of Exodus and their significance, thus becoming a focus for Israel's memory of its past, and its hopes for its future. When future generations asked why the Passover was being celebrated, they would be told about the deeper meaning of the event (Exodus 12:26–7). The festival of Passover was to be a permanent memorial of the mighty acts of God, which led to Israel being liberated from bondage in Egypt: 'Remember this day on which you came out from Egypt, out of the house of slavery, because the Lord brought you out from there by strength of hand' (Exodus 13:3). Once Israel had settled in the Promised Land, she was to continue the Passover ceremony as a way of remembering this act of divine deliverance.

Yet as the people of Israel wandered through the wilderness, they did more than remember the past; they looked forwards to the future. They looked *backwards* to the past, and recalled their period of captivity in Israel, and their liberation through Moses. Yet they also looked *forwards* to the final entry into the Promised Land, the eagerly awaited goal of their long journey. The uncertainties and difficulties of the present were thus sustained by the memory of past events and the hope of future events.

How We Fit Into Grand Narratives

I have already used the language of a 'big picture' in setting out the idea that there are certain frameworks of meaning – such as those offered by Marxism and Christianity – that provide an interpretation of the world and the place of human beings in it. Yet this is easily transposed into the idea of a 'grand narrative' – a story which has the imaginative power to change the way in which we see the world and ourselves. The term 'metanarrative' is often used to refer to such stories. Postmodern writers have been severely critical of these 'metanarratives', seeing them as authoritarian and normative. Yet ironically, they have simply created an alternative metanarrative, rather than abolishing the category altogether.

So how might we relate to these grand stories, such as those told by Marxism or Christianity? In both cases, the answer lies in our *understanding of the narrative* being supplemented by our *participation within the narrative.* Marx, for example, developed a narrative which affirmed the historical inevitability of socialism. He then invited his followers to become part of this narrative, and thus shorten the birth pangs of this process. They could work to hasten the coming of the revolution. Marx believed that what was historically inevitable could thus happen more quickly, as believers participated in – and thus accelerated the pace of – this grand narrative.

In the case of the Christian narrative, faith can be understood as a decision to become involved within this story. It is seen not simply as a means of making sense of the world, but as a way of becoming involved in the world. The New Testament uses a number of images to unfold the idea of individuals redirecting their personal stories, as they become part of something bigger and better. Paul's image of the crucifixion of the 'old Adam' is a good example. For Paul, Christians have been 'crucified with Christ', so that they now live 'by faith in the Son of God' (Galatians 2:20). Faith involves putting to death the old self, and rising to a new life. We do not lose our individuality; rather, we become *new* individuals. Our story is taken up within a greater story.

C.S. Lewis's *Chronicles of Narnia* can be seen as an imaginative retelling of the Christian 'big story' or 'grand narrative' of creation, fall, redemption and final consummation. One of the points Lewis makes in these works is that the stories of individual participants – such as Lucy Pevensie – are transformed, without losing their individuality, by becoming part of this bigger story. There is a sense in which faith is about embracing such a grand narrative, and becoming part of it.

The Christian sacraments also explore this point. The Eucharist – or Mass, or Lord's Supper – reminds Christians of the foundational story of their faith, and invites them to become part of this story. The sacraments thus recall the past, and at the same time invite believers to reshape their present in their light. The Christian story is something we are asked to enter and inhabit, not simply retell or recall. Meaning, identity and purpose are all safeguarded by becoming part of this grand narrative, and leaving behind a lesser, self-centred story. The quest for meaning is thus a search for the story in which we believe we are located, and working out what role we are to play within it.

Yet what if we simply make up our own story, creating an imaginary world of our own invention? How can we know that the quest for meaning is not illusory, something that is ultimately pointless? It's a good question, and we shall consider this in the next chapter.

6

Meaning: Discovery or Invention?

'It isn't just that I don't believe in God, and, naturally,
hope that I'm right in my belief. It's that I hope
there is no God! I don't want there to be a God;
I don't want the universe to be like that.'[1]

Thomas Nagel

The astronomer and cosmologist Carl Sagan (1934–96) once remarked that science 'counsels us to carry alternative hypotheses in our heads and see which ones best match the facts'.[2] Sagan and I seem to have ended up with a set of quite different working assumptions about life. But we both seem to agree that it is important to retain a sense of openness, a willingness to admit that we might be wrong. I like to think of this as a strength of character, rather than a weakness of intellect. It invites me to enter into the mindset of someone else, and try to understood what they believe and why – and to force myself to take their views seriously. And if we're going to carry these 'alternative hypotheses in our heads', we need to be exposed to these, and to be free to consider them.

Why We Need to Take Alternative Perspectives Seriously

As I have discovered from some rather depressingly sterile conversations, not everyone likes Sagan's idea. Dogmatic atheists and fundamentalist religious believers don't seem to like taking any intellectual possibilities other than their own seriously. People who are just intellectually lazy or who are aware of the vulnerability of their beliefs get round the problem by dismissing those with alternative viewpoints as deluded fools. To take the ideas of such people seriously, they argue, would simply contaminate their own intellectual and moral purity. We already know they must be wrong, so why waste time reading their ghastly books? They're little more than the infantile ramblings of stunted intellects.

The rest of us, of course, do not consider ourselves to be lesser human beings for failing to embrace these fundamentalist certainties, which generally turn out only to be 'certain' within their own ideological and cultural bubbles – bubbles that everyone knows will one day burst. Such fundamentalists already know that they are right, so they regard the exercise of even thinking about 'alternative hypotheses' as pointless and debasing. 'Don't insult me by asking me to consider such a ridiculous idea.'

Not surprisingly, both anti-religious and religious fundamentalists seem to have lists of 'safe' authors and books, and treat everyone else as heretical, deluded or just plain stupid.[3] Bullying rhetoric implies that anyone taking these ideas seriously is a traitor, a liar or a simpleton. 'I wouldn't be caught dead reading such rubbish!' is presented as if it were a moral or intellectual virtue, when it is little more than a rhetorical fig leaf covering an embarrassing ignorance, culpable lack of curiosity, or a suppressed intuition that my own ideas might be wrong. Yet within these bizarre fundamentalist bubbles, this vice is seen as a virtue. This, of course, is exactly how cultural 'in-groups' work. Both the New Atheists and religious fundamentalists tend to demonise anyone they regard as a threat. They may each use their own vocabulary, but the same basic strategy underlies them both: portray your opponents as mindless, mad or evil.

Sagan was right: it may be intellectually uncomfortable, but we need to respect other perspectives, and avoid demonising those ideas, or those who hold them. It makes polemics difficult. But that's how universities work. That's how science works. And that's how religion should work as well.

Let's look at science, and see why this matters so much. The phrase 'scientific orthodoxy' sends chills down my spine. Why? Because it has come to mean the dominant scientific view of an age, which self-importantly believes it is correct, and seeks to exclude alternative perspectives – for example, through the high degree of control exercised by peer-reviewed journals, which generally select articles favourable to the status quo. It is notoriously difficult to publish an article which questions a widely accepted theory, or presents data that is anomalous in terms of current understandings. Yet science works so well precisely because it changes its mind in response to evidence and theoretical development, which means abandoning what one generation believed to be 'orthodoxy'.

There are just too many examples of this for comfort. The Harvard psychologist Steven Pinker pointed out how many scientists routinely betray the core values of science in order to align themselves with 'morally progressive' ideas. In doing this, they abandon any pretensions to objectivity and neutrality, and become 'moral exhibitionists', denouncing their fellow scientists who fail to conform to their social and moral values.[4] Ideas are not evaluated in terms of their truth, but in terms of their consistency with the prevailing progressive ideas of racial and gender equality. Failure to conform to this orthodoxy means you aren't a member of the 'in-group'.

A good example of this cultural bias is found in the initial responses to Edward O. Wilson's classic *Sociobiology* (1975), which was widely interpreted on its publication by leading scientists and cultural commentators of the late 1970s as colluding with (if not actually openly advocating) racism, genocide and slavery.[5] Pressure mounted for these views (and Wilson himself) to be excluded from university campuses. That political posturing now lies in the past, even if other views are now demonised in this way; nevertheless, it is a disturbing reminder of how a groupthink can emerge, even within what many hope might be the most open-minded of disciplines: science.

In science, as in most other areas of life, certain 'alternative hypotheses' that ought to be kept open as genuine possibilities are rejected for social and political reasons, in that they do not conform to cultural norms – as if these were somehow arbiters of truth. It's all too easy to become a slave to intellectual fashion, letting transient cultural norms determine what's right through a process of alignment with individuals who are seen to embody cultural virtues of the moment. As the German poet and essayist Hermann Hesse pointed out, that was what happened in Germany in the 1920s. Everything was reduced to conformity to the latest intellectual fashions and 'transitory values of the day'.[6]

One of these 'alternative hypotheses' which I keep turning over in my own mind is this. *What if meaning is simply something that we invent?* In other words, it is something that I – or we – construct, reflecting my own personal interests, powerful voices that I feel the need to respect, or the views of the social groups to which I belong? Although I believe that meaning is something that we discern, can I entirely reject the idea that it is something that we make up?

It's an important question to ask. For a start, it raises the question of how someone can, with integrity, hold to a view which cannot be proved to be true, but which is nevertheless believed to be right. People regularly adopt a moral, social or political belief on the basis of some moral or social intuition, and then offer *post hoc* justifications for such views. The beliefs came first; only then were rational arguments made up to defend them.[7]

This raises deep questions about the boundaries of reliable human knowledge and the limits of human capacity to reason. Diehard rationalists who believe that you only accept ideas that can be proved to be right find themselves both frustrated and irritated by the human preoccupation with questions of meaning, which inevitably and necessarily transgress the limits of reason and evidence. The French philosopher Maurice Merleau-Ponty (1908–61) once remarked that, because human beings existed in the world, they were 'condemned to meaning'.[8] He seemed to be rather depressed by this thought. But that's just the way we are.

Given the importance of meaning to life, there is a clear and present danger that we might fabricate ideas to make ourselves feel better, or to conform to the 'group-think' which emerges within any human organisation – such as political parties, churches, the New Atheism, or groups of celebrities. How can we interrogate such ideas, to ensure we are not deluding ourselves? I am disturbed at how little today's defenders of the polity of the 'Age of Reason' fail to confront the fact that rationalism – as we find it, for example, in Nicolas de Condorcet – can serve a blatantly ideological function, in that it tries to justify the power of cultural élites over others.[9] It can easily become a form of narcissism, in which a self-proclaimed rationalist finds it 'difficult to believe that anyone who can think honestly and clearly will think differently from himself'.[10] This form of rationalism easily degenerates into a celebrity group-think, where an individual's cultural acceptance results from her publicly declared conformity to the ideas of a powerful individual.

Few people now read Howard Margolis's *Patterns, Thinking, and Cognition* (1987).[11] Yet its basic thesis must be noted by anyone reflecting on human nature, and especially human judgement – especially those who remain locked within the outmoded and discredited rational certainties of the 'Age of Reason'. Margolis noticed that people made political judgements, or came to political beliefs, on the basis of what is generally a tenuous or highly selective grasp of the evidence. It seemed as if human beings often provided rational justifications for beliefs that they had actually arrived at on other grounds. Above all, humans were very good at providing *post hoc* rationalisations of judgements or beliefs arrived at intuitively, not rationally.[12] The interconnection of human emotions and rationality is complicated, not least on account of the many neural sub-processes involved in human decision-making.[13] What makes a belief seem 'rational' requires far more nuancing than the philosophers of the 'Age of Reason' appreciated.

So how does this relate to the question of whether there really is such a thing as meaning in life? Let's begin by looking at the ideas of Ludwig Feuerbach (1804–72), a German philosopher whose ideas had a significant impact on Karl Marx and Sigmund Freud (whom we'll consider later in this work).

Believing What We Want? The Views of Ludwig Feuerbach

Why do people believe in God? The answer is far from simple. Richard Dawkins and other New Atheists treat God as some kind of object within the world, whose existence should be demonstrated by reason or scientific proof. Since no such proof is forthcoming, God is declared to be non-existent. Religious believers respond with a certain degree of exasperation to this unperceptive approach, pointing out that they do not think of God as a 'thing' or 'object' within the world. Rather, they take the view that there must be some fundamental agency or energy which is not conditioned by anything outside itself, if we are to make sense of the universe that we actually observe.[14] The English theologian William Inge, who later served as Dean of St Paul's Cathedral, London, put this point rather well:

> The real defect of rationalism or exclusive intellectualism lies in its attempt to prove Faith, or, I should rather say, in its belief that it has succeeded in demonstrating what cannot be demonstrated. Rationalism tries to find a place for God in its picture of the world. But God, "whose centre is everywhere and His circumference nowhere," cannot be fitted into a diagram. He is rather the canvas on which the picture is painted, or the frame in which it is set.[15]

Dawkins and Inge both agree on the importance of the rationality of belief systems. Yet they understand this notion in different ways. For Dawkins, a rational belief is one that is capable of being proved. Now this is an important view that needs to be taken seriously. Yet it too easily morphs into the view that reality is limited to, or defined by, what reason can prove. This turns out to be a rather narrow and constricted world. For Inge, the Christian faith provides a deeper understanding of rationality which acknowledges and explains the capacity (and limits) of human reason in the first place, offering a 'big picture' of our world which makes sense of its intelligibility and coherence to be understood. It is the eye of the needle, through which the world's threads are found to pass.

We shall return to consider issues of rationality later in this work. But let's return to that question about why people believe in God. What if we do so because we *need* to? Because we cannot bear the thought of a meaningless world? Because we are incapable of coping with the existential strain of living pointless lives in a pointless universe? Religious people will find this 'alternative hypothesis' unsettling. But it must be confronted.

Let's imagine someone explaining to us their grounds for belief in God. 'I hope there is a God! I don't want there to be no God; I don't want the universe to be like that.' The obvious danger here is that we simply invent the universe that we would like to inhabit. This objection to belief in God was given classic formulation in the writings of Feuerbach. God is an objectification of our deepest longings, the reification of our heart's desire, a projection of our feelings and aspirations onto an imaginary transcendent screen. There is no God. Our idea of God thus originates within us, and corresponds to nothing, save what lies within us. Although Feuerbach is usually regarded as an atheist, there is a sense in which he is actually an 'anthropotheist' – someone who believes in the divinity of humanity. Our concept of God does not arise from an external being, but from within us. We are the creators and origins of our own divinity.[16]

If Feuerbach is right, we ourselves create our notions of God. Given that there is no God (and we must emphasise that this presupposition is essential to the internal logic of Feuerbach's analysis), the idea can only come from within the depths of the human imagination. That's a challenging 'alternative hypothesis' for a religious believer, which was developed in different ways by Marx and Freud.

So what do I make of this? Well, I must take it seriously. A refusal to engage with this view would amount to an implicit admission that my worldview cannot cope with it. (That's one of the reasons why the New Atheism prefers to ridicule religion, rather than engage with its ideas seriously.) So let's begin to open this up for discussion.

First, I would agree with Feuerbach completely that nothing exists, nor needs to be true, because we feel it is true, or want it to be true. We need to come to terms with awkward yet verified truths, which some prefer to ignore, and others simply to deny. 'As an old philosophy lecturer once told us as fresh-eyed undergraduates in our first year at University: if something has been *proven* to be true, you just have to accept it, whether you like its implications or not. "It makes me feel bad, therefore it's false" is not an argument.'[17] I have no doubt that some would like there to be a God, and that this feeds into their reflections. Yet it's important to appreciate that others would equally like to live in a universe in which there is no God. I opened this chapter with a revealing admission from the philosopher Thomas Nagel: 'I hope there is no God! I don't want there to be a God; I don't want the universe to be like that.'[18] Nagel's philosophy is easily pilloried as a retrospective intellectual validation of a belief that had already been determined on emotional grounds – a good example of what the psychologist Jonathan Haidt describes as the 'emotional tail' wagging the 'rationalist dog'.[19] Nagel offers a *post hoc* intellectual justification of beliefs that merely express what he would like to be true. And he's not on his own here.

Many people do not want there to be a God, so that they can be the masters of their own destinies without having to worry about any divine interference. C.S. Lewis is a classic example of someone whose early atheism reflected his desire for autonomy, safe from being troubled by the 'Great Interferer'.[20] Historians often point out how the origins of modern atheism reflect a longing for humans to be able to do anything they want, rather than work within a divinely given moral framework.[21] A desire for total human autonomy was expanded into a desire for the elimination of perceived obstacles to such freedom – including the notion of God.

Feuerbach's 'given' is that there is no God. Feuerbach does not prove this core belief, nor offer any warrants for believing that it is right. Yet this core assumption becomes the driver for his engagement with religion. Since there is no God – Feuerbach's fundamental axiom – it follows that the observation that many people believe in God implies that they are misguided or deluded.

A Christian 'given' is that there is a God, and that human existence originates from and finds its fulfilment in God. As Augustine of Hippo put this, using the framework of a prayer to God: 'You have made us for yourself, and our heart is restless until it finds its rest in you.'[22] This theological framework immediately suggests the entanglement of desire and reason, offering a framework which explains why we long in the first place, and what we really long for in the second. Like Feuerbach's approach, this proves nothing, save the internal consistency of the framework itself.

Both Augustine and Feuerbach actually offer a 'logic of desire' – that is, a rational explanation of our sense of longing, how it arises, and what it intimates. Yet they understand the nature, origins and goals of this desire in quite different manners. For Augustine, God is there to be discovered and encountered; for Feuerbach, there is no God, and some compensate for this apparent deficiency by inventing one to meet their own expectations or desires.

What if Feuerbach is right? What if there is no transcendent realm as such, but only a set of ideas or values which we ourselves have constructed, and choose to treat as if they are transcendent? What would it be like to inhabit such a world?

Making Up Meaning

Feuerbach's elimination of a transcendent realm from thinking about religion, morality and meaning raises the question of the origins of these notions. If Feuerbach is right – and that, by the way, is a massive 'if' – then our ideas about values and meaning must be recognised as purely human constructions, without any transcendent ground or warrant. Humanity creates its own values and ideas, and is not accountable to any external objectivity in doing so, precisely because there is no such transcendent reality that need be acknowledged or referenced.

That was certainly the view of the German philosopher Friedrich Nietzsche, who was convinced that the world had no intrinsic meaning. We could either learn to live with meaningless-ness, or invent our own meaning and impose it on the world and our lives. Nietzsche argued that the collapse of any belief in an objective truth about the world was liberating, in that it set human beings free to create their own ideas of meaning and value.[23] For Nietzsche, there are no objective facts; just subjective opinions.

This fashionable (though deeply problematic) view is widely affirmed within postmodern philosophy. For Jacques Derrida, the notion of *la différance* expresses the ultimate undecidability of all meaning. Resolution of the question of meaning is endlessly deferred and is therefore never fixed.[24] A similar view is expressed by the philosopher Richard Rorty, who teased out its inevitable implications. 'There is nothing deep down inside us except what we have put there ourselves, no criterion that we have not created in the course of creating a practice, no standard of rationality that is not an appeal to such a criterion, no rigorous argumenta-tion that is not obedience to our own conventions.'[25] We *invent* meaning – including our ideas of identity, purpose and value.

Some people will find this idea to be unsettling, where others may find it liberating. Yet it is only an intellectual possibility. Our emotive reaction to this possibility is no indicator of its truth, although it will inevitably shape its reception. We can be like Feuerbach, and make subjective emotional resonance a criterion of truth. 'I like this – therefore it's right.' It's a highly narcissistic view that appeals to those who see themselves as arbiters of truth, as well as fashion. Most, however, are likely to feel that such a self-serving and self-justifying view just can't be right. So what possible responses might be developed to Rorty's suggestion that we invent meaning and morality to conform to our expectations and aspirations? After all, there is ample evidence that people do this all the time.

A number of points could be made. To begin with, we may note that Rorty's position is vulnerable, in that it seems to do little more than reify social practices, and treat these as being synonymous with 'truth', 'goodness' or 'justice'.[26] For Rorty, what is right, good or meaningful is determined by what we *do* as individuals or groups. And it is obvious that what people do changes over time. Social norms are notoriously volatile and unpredictable, and, as the Marxist theoretician Antonio Gramsci so clearly demonstrated, are easily manipulated by those in power or with vested interests.[27] Rorty thus locks himself into a shifting and unstable world, within which no ultimate resolution of the question of meaning is possible. The question may *seem* to be resolved, yet in reality this is simply a social consensus created and manipulated by those in positions of cultural power and authority.

The fundamental problem seems to be that Rorty's approach is not capable of offering a criterion that stands *above* human practice, by which rival visions of meaning or morality could be judged.[28] If we invent meaning, then we need to evaluate competitive concepts of truth, goodness and purpose in life. Yet if there is no external or objective vantage point from which this can be done, we can only make those judgements on the basis of subjective perceptions or prevailing social consensus within the group that we consider to be normative.

In effect, this is the dilemma which is embraced as a virtue by postmodernity, which declines to accept the notion of any normative standpoint or objective perspective by which ideas can be judged. Yet the downside of this approach can hardly be overlooked. What defence mechanisms does it propose to guard against delusion, intellectual manipulation by power groups and vested interests within society, and sheer personal dishonesty? As

Alasdair MacIntyre pointed out, all too often 'what purport to be objectively grounded claims function in fact as expressions of arbitrary, but disguised, will and preference'.[29] Postmodernity's failure to offer persuasive answers to these questions has led many to believe that it 'has died in a kind of *fin de siècle* despair at its inability to interrogate the consequences of its own provisionality and indecipherability'.[30]

If we invent notions of meaning, they still require evaluation. The essentially political judgement that all such notions are of equal value or plausibility experiences serious discomfort when challenged on the grounds of its intellectual accountability. This difficulty is often met through the invention of criteria – that is, the formulation of rules by which beliefs may be judged. Yet those criteria ultimately reflect the consensus and practices of a community. They are socially constructed, not embedded within some deeper order of things, permitting them to have an objectivity and validity independent of the thinker.[31] Criteria of evaluation are conventions that arise within a community. They are not etched into the glass of the 'mirror of nature'. And sometimes those criteria are devised precisely to secure an intended outcome. What matters is securing a specific outcome; the criteria are invented with this objective in mind.

This rather leaves us floundering. If we follow these lines of thought through to their brutal end (something which understandably few choose to do), we find ourselves in an intellectual quagmire, being asked to assess invented meanings using invented criteria. Most draw back from this on grounds that are actually intuitive or emotional, rather than rational. We don't *want* this to be the case; or we *feel* it can't be right. Yet there is no evading the outcome of any theory of meaning that locates its origins and grounds in individual or communal subjectivity. There is, of course, a way of challenging this – but it is one that calls into question the fundamental presuppositions of this whole approach to meaning. It is to consider that there might be some objective grounds for meaning – for example, in questions of values and purposes.

This was the point made throughout her philosophical writings and novels by Iris Murdoch. For Murdoch, notions of meaning and goodness can only be stabilised through being grounded in some transcendent realm or framework.[32] Goodness is not something that we *invent*; it is something we *discover*. In the light of this transcendent good, we come to recognise the deficiency of our own compromised ideas of morality, which are too often self-serving. In one sense, the basis for true morality is 'un-selfing', a dying to self in which we become attentive to the calls and demands of 'the good', which lies beyond and above us.[33] We are thus accountable for our notions of meaning, in that they are grounded in something that lies beyond us, that is not under our control, which acts both as the basis for and judge of our understandings of meaning.

Murdoch's emphasis on the importance of the transcendent in stabilising moral notions was anticipated by the great Russian novelist Fyodor Dostoyevsky in his novel *The Devils* (1871–2), widely regarded as one of the greatest literary criticisms of nihilism. The most important character in the novel is Kirillov, who argues that the non-existence of God legitimates all forms of actions. 'If God exists, then everything is His will, and I can do nothing of my own apart from His will. If there's no God, then everything is my will, and I'm bound to express my self-will.'[34] Since the idea of God is a pure human invention or construction, Kirillov reasons that he is therefore free to do as he pleases. There is no higher authority, to whom he is ultimately accountable, or who is able to negate his totalitarian moral self-assertion. There is no defensible alternative to an ethic of self-actualisation.

Dostoyevsky comes close to the view famously expressed by Glaucon, the cynic in Plato's *Republic*, who told Socrates that people would behave ethically only if they thought they were being watched. Dostoyevsky takes much the same view in his 1878 letter to Nikolai Ozmidov, in which he sets out the implications of atheism for morality:

Now assume that there is no God, or immortality of the soul. Now tell me, why should I live righteously and do good deeds, if I am to die entirely on earth? ... And if that is so, why shouldn't I (as long as I can rely on my cleverness and agility to avoid being caught by the law) cut another man's throat, rob and steal?[35]

The idea of God watching you, seeing what others overlook, clearly has implications for our sense of accountability!

So do we have to make a choice here? Do we have to choose between an objective concept of meaning, discerned or intuited from the world around us, or a subjective concept of meaning, which arises within us? Or might there be a way of holding these together coherently?

Objective and Subjective Approaches to Meaning

I want to suggest that meaning is something that is *objectively grounded* but is *subjectively applied*. I agree with Murdoch that we cannot just make things up; we want to feel that we respond to something that we believe to be outside us. Yet *an objective truth needs to be subjectively appropriated*. We need to embrace and apply our understanding of meaning, so that it impacts on our subjective experience. One of the most influential discussions of this point comes from the philosopher Søren Kierkegaard, who argues that truth was something that impacted upon us deeply and *inwardly*. It is not enough to accept that something is true or meaningful in a distant and detached manner; this meaning or truth has to be embodied in our actual lives and subjective experience. In short, he declares, truth or meaning demands 'an appropriation-process of the most passionate inwardness' if it is to become a reality within the life of an individual.[36]

We have already noted the views of the philosopher Susan Wolf, who points out that 'academic philosophers do not talk much about meaningfulness in life'.[37] Wolf singles out two alternative approaches to securing meaningfulness in life.

The first view is the need to 'get involved in something "larger than oneself" ' to secure meaningfulness.[38] This view affirms or assumes the importance of 'participating in or contributing to something whose value is independent of oneself'.[39] This objective approach involves the recognition of the importance of something beyond the individual thinker, which is held to be meaningful in itself.

A second view holds that 'it doesn't matter what you do with your life as long as it is something you love'.[40] This subjective view encourages the pursuit of what we feel to be meaningful, and judges their validity by 'the subjective quality' of the 'inner lives' of those who follow this course.[41]

Some may feel that this second approach risks collapsing into what C.S. Lewis termed the 'poison of subjectivism'.[42] It is, he remarked, a 'fatal superstition' to believe that human beings create values simply as a matter of taste, as someone might choose their clothes.[43] Yet while Lewis was severely critical of self-indulgent wallowing in an experiential quagmire, he firmly believed that the quest for meaning pointed to something beyond us that nevertheless evoked a powerful and appropriate subjective response within us. Like Kierkegaard,[44] Lewis held that encounter with truth necessarily changes our inner existential worlds.

Yet Wolf's two views can clearly be seen as complementary, rather than as competitive. Suppose that I recognise something as intrinsically valuable in itself. Why could I not then make a subjective commitment to love this, and make its pursuit my goal in life? This is what I found happening to me, as I moved away from atheism towards Christianity. My initial attraction to Christianity was intellectual – a sense that it offered a fuller, richer and more satisfying account of the world than atheism. Although atheism then seemed to me to be emotionally sterile, I chose to frame this as a virtue, seeing its sheer bleakness and austerity as potentially an indication of its truth. At that stage in life, my focus was purely intellectual. When I came to the conclusion that Christianity made more sense of life than atheism, I had no sense that Christianity might turn out to have emotional, imaginative or aesthetic dimensions – or that these might be important.

Like Dorothy L. Sayers before me, I fell in love, not with the person of God, but with the pattern of Christian truth. It was an *eros* of the mind, not the heart. Just as Sayers suspected she had 'fallen in love with an intellectual pattern',[45] so I found myself drawn to what I was realising to be the explanatory capaciousness of the Christian faith.

Although my journey to faith began with an objective appreciation of the intellectual merits of faith, it most certainly did not stop there. As I began to explore what I had embraced, I slowly came to appreciate the relational, aesthetic, emotional and imaginative aspects of faith. These were absent from my atheist worldview, and I had naturally assumed that they would also be absent from Christianity. Slowly, the pieces of the jigsaw puzzle began to fit together. God remained an explanatory principle for me then, as now. Yet I gradually grasped the importance of the characteristically Christian notion of a *personal* God as someone who could be known, not simply known about, and whose relationship with me could be characterised in terms of love, not simply intellectual illumination. The objective and subjective sides of my faith began to coalesce into a greater whole.

This may have been new to me; it was, however, a standard and unproblematic position for many Christian theologians. I later found this approach explored with imaginative power in C.S. Lewis's *Pilgrim's Regress* (1933), which affirmed the need to hold subjective and objective aspects of faith together, avoiding purely cerebral or emotional accounts of faith. A balancing act is certainly required – but Lewis himself believed that it was possible to construct a *via media*, a 'middle way' which embraced and enfolded both the objective and subjective aspects of meaning.

Relationships and Meaning

Recent psychological research has confirmed the importance of personal relationships in creating and sustaining a sense of meaning in life. Critically important terms such as 'love', 'trust' and 'commitment' naturally belong within the context of a personal relationship. People often define who they are and why they matter in terms of relationships. The deep human need to love and know that we are loved naturally expresses itself in a variety of relationships. In his influential work *The Four Loves* (1960),[46] C.S. Lewis noted four broad categories of personal commitment – affection, friendship, *eros* and charity – which give human life its meaning and value. While these relational categories may turn out to have no 'survival value', Lewis remarks, they are rather something which gives 'value to survival'.[47] Loving someone – and being loved in return – makes life meaningful and worthwhile.

A relationship bridges our objective and subjective worlds. This point is often developed using the language of the Jewish philosopher Martin Buber. In his classic work *I and Thou* (1922), Buber drew a distinction between two quite different kinds of relationship: 'I–It' and 'I–Thou'. An 'I–It' relationship is that between an active subject and a passive object. It is the sort of relationship we might have with a pen, a stone or a piece of paper. We treat it as a passive object, having a certain colour, weight or size. We know *about* a piece of paper in an impersonal, objective way.

An 'I–Thou' relationship is totally different. It is about two active subjects, who freely enter into a relationship. Two lovers might know a lot about each other; they certainly *know* each other – personally and subjectively. Their relationship is an invisible bond between them, which changes dramatically the way in which they see each other, and the world around them.

Buber's theory is more than just a way of helping us to make sense of relationships. Buber was critical of the depersonalising trends that rose in the aftermath of the First World War, and became particularly significant in the 1930s. The Soviet Union took a collectivist approach to life, which seemed to eliminate personal identity in order to serve the needs of the State.[48] Buber's approach helped clarify what happened in this dehumanising process, initially in the Soviet Union, and then in Nazi Germany. In the West, capitalism commodified humanity, calling into question the significance and value of the human individual beings.[49] An individual human was treated as an 'It', not a 'Thou' – as a depersonalised body, whose significance lay only in its capacity to work for the State. Our importance was functional, not relational. This also allows us insights into the theology of sin, which can be understood as a relational notion, expressed in dysfunctional relationships with other people, the environment and God.

Yet Buber himself was clear that his approach did more than safeguard human identity; it also helped illuminate what was for him one of the most important dimensions of human nature – the human relationship with God. For Buber, God was the 'eternal Thou', who entered into a transforming relationship with a human 'Thou'. God was not to be thought of as an abstract, impersonal force, but as a living reality who could be known, trusted and loved. To relate to God is to be loved and affirmed, and transformed in and through this new relationship, in which we become more like God precisely because we relate to God. For Lewis, we are 'mirrors' whose brightness is 'wholly derived from the sun that shines upon us'.[50]

Meaning, purpose and value are strongly subjective ideas, which matter profoundly to people, and are probably best expressed in terms of a relationship. Yet these qualities are not merely subjective; they are vulnerable. So what happens if we lose any sense of meaning or purpose? We shall explore this challenging theme in the following chapter.

7

When Meaning Fails:
Doubt, Trauma and Disbelief

'It's really a wonder that I haven't dropped all my ideals,
because they seem so absurd and impossible to carry
out. Yet I keep them, because in spite of everything,
I still believe that people are really good at heart.
I simply can't build up my hopes on a foundation
consisting of confusion, misery, and death.'[1]
Anne Frank

Like any scholar, I am passionately committed to the pursuit of truth, irrespective of whether I like what I find or not. I share the view of the great English philosopher John Locke, who wrote of the sense of fulfilment, even delight, that accompanied the search for truth in life:[2] 'I know there is truth opposite to falsehood, that it may be found if people will, and is worth the seeking, and is not only the most valuable, but the pleasantest thing in the world.' Yet anyone who pursues truth soon discovers that it is frustratingly elusive, lying beyond our grasp. We can, of course, prove certain truths in logic and mathematics – such as Gödel's incompleteness theorem, which neatly subverts the idea that all propositions that *are* true can be *shown* to be true. Yet the truths that give meaning and purpose to life lie beyond rational or empirical demonstration.

Some more exuberant writers of the Enlightenment held that truth could be permanently and clearly established, independently of the thinker's cultural and historical situation. Benjamin Franklin famously declared his belief that certain truths are 'self-evident' – most famously, in his statement in the 'Declaration of Independence' that 'all men are created equal, that they are endowed by their Creator with certain unalienable Rights, that among these are Life, Liberty and the pursuit of Happiness'. (Thomas Jefferson's original line in the 'Declaration of Independence' – which was edited by Franklin – spoke instead of holding 'these truths to be sacred and undeniable'.) Yet this is not an evidenced argument. A 'self-evident' truth is basically an intuition, in which someone just 'sees' that something is right, without the need for evidence or argument.[3] And many of those 'intuitions' turn out to be shaped by our cultural location.

In his *Dark Materials* trilogy (1995–2000), Philip Pullman introduced the idea of the alethiometer – a mechanised device which determines truth algorithmically and objectively. It's a great literary expression of the Enlightenment's dream of an objective truth which lay beyond the contingencies of history and human cognitive bias. Yet Pullman's narrative in the *Dark Materials* trilogy shows that the *interpretation* of the alethiometer's outputs is problematic, save apparently for Lyra, the central character of the trilogy, who seems to have an intuitive grasp of the meaning of the device's symbols.

The alethiometer is a symbol of the Enlightenment's doomed quest for an objective truth, binding on all rational beings, which could be established through a mechanical process which was independent of history and culture. Yet we now know that the rationalising algorithms of Descartes and Spinoza did not – and could not – answer life's greatest questions. Although the problems of this approach were obvious from the outset, it was only in the aftermath of the First World War that a serious critique of its plausibility emerged in the writings of Wittgenstein and Heidegger.[4] Perhaps we are now more willing than before to recognise the epistemic limits of humanity, in which we realise that we have to trust, rather than expect to prove, our core beliefs.

As Bertrand Russell remarked, philosophy tries to teach us 'how to live without certainty, and yet without being paralyzed by hesitation'.[5] Religion is the supreme cultural exemplar of a chastised and realistic account of human rationality, not its antithesis.

It affirms the importance of reason, while recognising its limits, thus avoiding the delusional tendencies of earlier generations to idealise reason, detach it from human emotions and feelings, and fail to appreciate how easily it can go wrong. The contemporary resurgence of interest in classic philosophical theologians such as Augustine of Hippo and Thomas Aquinas reflects many considerations, one of which is that they represent an approach to reason, experience and imagination that exceeds both in elegance and utility the thin mechanical philosophy of the Enlightenment.

Despite our divergences in belief, both Richard Dawkins and I are trapped within the human condition of being unable to prove what we believe. I believe there is a God; Dawkins believes that there is no God. This does not mean that Dawkins and I have arbitrary beliefs; it just means we cannot prove them to be true, and have to learn to live with this tension – not because we enjoy this, or get some kind of weird existential satisfaction from hovering on the borderlands of a tantalisingly distant truth, but because of the limits placed on human beings to discern truth in the first place. We feel it is important to grasp deep truths, but we discover that it is difficult to do so.

There is nothing new or controversial about this. The intellectual historian Sir Isaiah Berlin (1909–97), noted for his work on ideologies, pointed out that human convictions fall into three basic categories:[6] those that can be established by empirical observation; those that can be established by logical deduction; and those that cannot be proved in either of these ways. The first two categories concern what can be known reliably through the natural sciences on the one hand, and what can be proved through logic and mathematics on the other. The third category concerns the values and ideas that have shaped human culture and given human existence direction and purpose; these, however, cannot be proved by reason or science.[7] Examples include – but are not limited to – moral, political and religious beliefs.

Even scientific truths are provisional, in that they represent the most reliable judgements of the scientific community at the present moment. Yet these change over time, as new evidence emerges. Back in 1900, the scientific consensus, based on classical physics, was that the universe had existed for ever. Two reasons in particular should be noted. First, Newtonian physics held that time was infinite in both directions, precluding any notion of the 'beginning of time'. Second, it was thought that the principle of the conservation of energy required that the universe had to be of infinite age, in that the notion of a cosmic beginning seemed to violate this principle.

Today, we believe the universe came into existence in a singular event that we call the 'Big Bang'. What will the scientific community believe in 2300? We do not know, and have no means of knowing. Some of today's core scientific beliefs will unquestionably be preserved; others, however, will be abandoned, in that they will no longer be seen as the best way of making sense of the evidence available. The difficulty lies in knowing which will be retained, and which left to wither by the wayside.

Science here mirrors a wider realisation of the need to live with uncertainty. The world is complex, not easily reduced to the certainties of any philosophy or theology. Pascal's reflections on the opaqueness of the world is not a mark of scepticism. Though firmly committed to believing in a loving and benevolent God, Pascal had no doubt that nature witnessed ambivalently, if at all, to such a divinity.

> I look on all sides, and I see only darkness everywhere. Nature presents to me nothing which is not matter of doubt and concern. If I saw nothing there which revealed a Divinity, I would come to a negative conclusion; if I saw everywhere the signs of a Creator, I would remain peacefully in faith. But, seeing too much to deny and too little to be sure, I am in a state to be pitied.[8]

If it is valid, Pascal's insight subverts the wisdom of our age, in which simple slogans triumph over nuanced attempts to do justice to the complexity of our world, and human attempts to understand it. Why, to use a courtroom analogy, do so many recent philosophers sound like lawyers trying to build their case by multiple technical arguments, rather than like judges who deliver revelatory verdicts of truth? Pascal's point is that simple worldviews require the cultivation of blindness on the part of their advocates. It is only by being deliberately inattentive to some aspects of our universe that we can construct the kind of totalising 'grand narrative' that we find in secular faiths such as Marxism or the rationalist platitudes of the 'New Atheism'.

There must be room for doubt and uncertainty in life, precisely because neither the universe around us nor our minds within us yield the certain deliverances that are required by fundamentalisms, whether religious or anti-religious. What Karl Popper called 'ultimate questions' elude our intellectual grasp and lie beyond strict logical proof. The problem is not faith, which is a natural and inevitable part of human life, but a shallow and glib dogmatism which refuses to recognise the human need for faith in the first place.

Doubt is just part of the intellectual and emotional routine of life. There is a sense in which we are all agnostics, in some sense and to some degree. We can, of course, prove shallow truths, yet the beliefs in life that really matter lie beyond proof. There may be good reasons for thinking that certain beliefs are right. But in the end, we just cannot demonstrate that this is the case. As Bertrand Russell pointed out, that is just the way things are, and we need to learn how to live without the certainty that we crave.[9]

These points are not controversial; they are, however, deeply unsettling to those who are attracted to worldviews because they believe they offer them certainty. Some are drawn to intensely dogmatic forms of religion for this reason; others to the superficial yet equally dogmatic platitudes of the New Atheism. Most of us, however, have got used to the fact that leading meaningful lives demands stepping beyond the secure deliverances of reason – without thereby degenerating into some kind of irrationalism. If we are to cope with life (and not merely make sense of it), we need answers to the question 'why', not merely to the question 'how'. Let's look at this point in more detail.

Making Sense of Things – Coping with Things

How do we cope with life? In his remarkable book *The Twilight of the Idols* – with the intriguing subtitle 'How to Philosophize with a Hammer' – the philosopher Friedrich Nietzsche answered this question with a powerful one-liner. 'If someone knows the "why" of life, then the "how" can look after itself.'[10] If you have found the 'why' of life, according to Nietzsche, then you can cope with just about any 'how'. Nietzsche's book reflects his deep disillusionment with the academic philosophy of his day. Why did so many philosophers think that life was worthless? Nietzsche's criticism of Socrates may be wide of the mark, but his point needs to be taken seriously. We need a philosophy of life which is able to engage positively and practically with the realities of everyday life, rather than retreating into abstract ideas or a preoccupation with heaven that distances us from the cares and concerns of this book.

Nietzsche's aphorism neatly anticipates a core theme of much recent psychological research: it is much easier to cope with life if we can discern purpose and meaning within events, even when these are traumatic. Empirical studies of many groups dealing with major life stressors – such as natural disasters, illness, or loss of loved ones – show that religion and spirituality are usually found to be helpful to people as they try to cope with these situations. This theme was explored some time ago by Viktor Frankl (1905–97), whose experiences in Nazi concentration camps during the Second World War helped him to appreciate the importance of discerning meaning in coping with traumatic situations.[11] Frankl argued that survival in concentration camps depended on maintaining the will to live, which involved the discernment of meaning and purpose even in highly demoralising situations. Those who coped best with these appalling conditions were those who had frameworks of meaning that enabled them to fit their experiences into their mental maps of meaning.

More recently, the psychologist Kenneth Pargament has argued that religious faith provides a framework within which responses to crisis may be developed. Religious faith offers resources through its teachings, rituals and narratives for both 'religious coping through conservation' (that is, assimilating these events into one's faith perspective and enriching it) and 'religious coping through transformation' (that is, accommodating one's faith perspective to these events, so that it is challenged and revised in their light).[12]

There is now a growing consensus that religion and spirituality are of major importance in coping with stressful situations, such as natural disasters – for example, by providing a map of meaning which allows individuals and communities to re-orientate their subjective perceptions of situations.[13] These frameworks of meaning help them to make sense of what happened, cope with the stress it generates, and move on from there.

It is important to draw a distinction between *making sense of a situation* and *coping with it*. Let's go back to our earlier analogy of the Balcony and the Road. Someone observing a traumatic event from the safe distance of the Balcony will try and make sense of what is happening. It may be intellectually interesting, but the observer is detached and disengaged. The situation faced by people on the Road is very different. They are participants, as much as observers, being caught up in the events that they are experiencing. Their need is not simply to make sense of things, but to cope with it – to survive, and perhaps even to become stronger and better people as a result of their experience. The Balcony yields an objective account of things, leading to comprehension; the Road involves a subjective immersion in experience, which needs – and hopefully generates – resilience.

Faith can help us cope with trauma. Yet there is another side to this. What if trauma calls faith itself into question? What if that map of meaning provided by religious faith is overwhelmed by events and experience? The erosion or loss of a map of meaning is not, of course, limited to religious faith. Arthur Koestler enthusiastically embraced Marxism-Leninism in 1931, yet his political faith was placed under stress by the rise of Stalinism, which called into question many of his core assumptions. Finally, Koestler realised he could no longer maintain his position with integrity. He resigned from the Communist Party in 1938.

In his autobiography, Koestler describes his own gradual movement away from his youthful certainties about how a Marxist framework made perfect sense of the world to a reluctant recognition of its obscurity and resistance to definitive interpretation. 'In my youth, I regarded the universe as an open book, printed in the language of physical equations and social determinants, whereas it now appears to me as a text written in invisible ink, of which, in our rare moments of grace, we are able to decipher a small fragment.'[14] So what causes this kind of collapse of meaning in the face of trauma? There have been many psychological studies of the impact of trauma, which emphasise the threat that it poses not only to the physical survival of the individual, but also to core positive beliefs about the world or the self, especially through the shattering of core personal assumptions relating to meaning and self-worth, or the confirmation of converse negative assumptions.[15]

People often rely on what may seem to be somewhat naïve assumptions to enable them to cope with life. The case of Anne Frank is especially significant. Frank was a member of a Jewish family who lived in hiding in Amsterdam during the Nazi occupation. After their hiding place was discovered in August 1944, Anne was transferred to Bergen-Belsen concentration camp, where she is believed to have died in February 1945 at the age of fifteen. In a diary entry for June 1944,[16] Anne wrote of her abiding faith in the goodness of human nature, which she felt unable to abandon, despite everything around her which called it into question as 'absurd and impossible'. To abandon this belief would amount to an admission that life was based on nothing more than 'confusion, misery, and death'. It was clearly a conclusion that

Anne found unbearable. We have no way of knowing how Anne's experience at the Bergen-Belsen concentration camp influenced her determination to continue believing that 'people are really good at heart'. It seems empty and totally inadequate to suggest that her experience would have been traumatic.

Human evil is one of the most significant causes of a catastrophic questioning of faith; another is the experience of intense and seemingly pointless suffering. One of the most remarkable and challenging accounts of the impact of such trauma on faith is found in a late work by C.S. Lewis. In the next section, we shall consider Lewis's changing views on suffering, and reflect on how they illuminate the issues we're considering.

C.S. Lewis on Suffering

We have already noted how Lewis is a 'big picture' thinker, committed to the idea that the Christian faith offered a better imaginative rendering of reality than his earlier atheism. Lewis's distinct way of approaching Christianity involves both the reason and imagination, and is best thought of in terms of a recognition of the capacity of faith to accommodate what we observe around us and experience within us.[17] So how does his way of thinking 'fit in' suffering?

Lewis's first book of Christian apologetics, *The Problem of Pain*, appeared in 1940, and was well received. It was an attempt to show how the existence of suffering could be accommodated convincingly within a Christian way of thinking. Its most famous one-liner is Lewis's declaration that pain is God's 'megaphone to rouse a deaf world'.[18] The work's clarity and accessibility endeared it to many, and laid the foundation for Lewis being invited to give a series of 'broadcast talks' by the British Broadcasting Corporation, later published in a slightly revised form as the phenomenally successful *Mere Christianity* (1952).

The Problem of Pain remains a valuable exploration of the intellectual issues raised by suffering and pain, and many continue to find its approach helpful and plausible. Yet Lewis's reconciliation of faith and experience is rational, not existential. The work focuses on the interconnection of abstract ideas, not the distress and anxiety caused to people by the harsh, brutal realities of suffering and death.

Many readers of *The Problem of Pain* have found themselves wondering whether there is a near-total disconnection between Lewis's intellect and his emotions. Lewis himself seems to have realised this. In a letter to his brother Warnie, written while he was working on the book, Lewis appeared to suggest that the experiences of 'actual life' have no bearing on the essentially intellectual issue under discussion. What has the experience of suffering got to do with explaining how it fits into the perspective of faith?[19] Lewis here seems to imply that intellectual reflection is detached from the world of experience, which has little relevance for such abstract discussions. So what would happen if Lewis were himself to experience overwhelming distress in the face of pain – either within himself or through someone whom he loved, and whose pain he felt as his own?

In 1961, a short work by N.W. Clerk appeared, with the title *A Grief Observed*. It consisted of the painful and brutally honest reflections of a man whose wife has died, slowly and in pain, from cancer. Those reflections include a vivid depiction of his own reaction to her death, as well as some more theological reflections on the goodness of God. How can what has happened make sense, if God is good and loving? Clerk writes of his realisation that his rational, cerebral faith has taken something of a battering from the emotional crisis that has overwhelmed him. The ideas that had once proved intellectual anchors to his life turned out to be inadequate in the face of such a catastrophe.

We now know that 'N.W. Clerk' was actually a pseudonym for none other than C.S. Lewis. There is a sense in which *The Problem of Pain* laid the groundwork for the emotional firestorm of *A Grief Observed*. To its critics, Lewis's approach in *The Problem of Pain* amounts to an evasion (or even denial) of pain and suffering as *experienced realities*. Instead, they seem to be reduced to abstract ideas, pieces which need to be fitted into the rational jigsaw puzzle of faith. To read *A Grief Observed* is to realise how a rational faith can fall to pieces when it is confronted with suffering as a personal reality, rather than as an inconvenient theoretical disturbance. *A Grief Observed* is saturated with poignant, heart-rending cries of despair and bewilderment, as Lewis's maps of meaning seem to fall apart in the face of the unrelenting pain of loss. 'Where is God? Go to him when your need is desperate, when all other help is vain, and what do you find? A door slammed in your face, and a sound of bolting and double-bolting on the inside. After that, silence.'[20]

Lewis's powerful, frank and honest account of his own experience is to be valued as an authentic and moving account of the impact of bereavement. It is little wonder that the work has secured such a wide readership, given its accurate description of the emotional turmoil that results from a loved one's death. Yet the work is significant at another level, in exposing the vulnerability and fragility of a rational faith that is rooted only in the mind.

Lewis's correspondence in his final years of life shows that he fits into the pattern now generally known as 'post-traumatic growth' – that is, not a mere return to the baseline of faith before the traumatic event, but rather an advance or enhancement, in which an individual comes to find deeper levels of meaning and of personal significance than before the event.[21] Earlier we noted Kenneth Pargament's important distinction between 'religious coping through conservation' (in which experiences are incorporated into someone's faith perspective and come to enrich it) and 'religious coping through transformation' (in which experiences cause an accommodation of someone's faith perspective to these events, so that it is challenged and revised in their light). So which of these better describes what happened to Lewis?

We can't say for certain, partly because Lewis did not live long after the publication of *A Grief Observed*, and did not write about his experience of suffering in sufficient reflective detail to allow us to answer this question definitively. Yet Lewis's experience helps us to explore in more detail the important distinction between making *sense of suffering* and *coping with suffering*. Let's reflect more on this.

The Role of Faith in Sense-Making and Creating Resilience

I have taught theology at Oxford University for many years, and regularly lead discussions with students on various topics. One of those focused on what are traditionally known as 'theodicies' – ways of making sense of suffering from a Christian perspective. These are often highly abstract intellectual explanations of why the existence of pain in the world does not defeat belief in a good God. Students enjoy kicking around the ideas, and opening up the ideas of writers such as Augustine, Leibnitz and John Hick.[22] I hope I don't sound cynical, but it sometimes seemed to me like shadow-boxing – discussing ideas that are reasonably interesting, but don't actually make any difference to those discussing them.

Another student discussion I led around the same time focused on the issue of suffering in Christian spirituality. Here, the agenda was very different. We were looking at how the great spiritual writers of the Christian tradition dealt with the experience of suffering. How could faith uphold us as we suffer? How could it help us become better people? How could these spiritual ideas and approaches help us to put our lives back together again, and cope with suffering, while at the same time aspiring to learn from it? The mood of this seminar was quite different. Students were looking for help and guidance, seeing the authors under consideration as sources of wisdom which might change their lives for the better.

The marked difference between these two seminars reflects a fundamental divergence between academic theology and Christian spirituality – between a mindset that wants a clear, truthful and objective account of the Christian faith, and another that seeks inner personal transformation, leading to a greater resilience in the face of stress and trauma. The two can, of course, be bridged in helpful and important ways,[23] yet all too often they seem to be treated as two separate watertight compartments within the Christian mind. Historians will, of course, rightly point out that this bifurcation of theology and spirituality is quite recent, and is capable of being reversed. But there's a lot more work that needs to be done to make this happen.

Where academic theologians explore the cognitive tensions arising from suffering in an objective and detached manner, most spiritual writers focus on how people could deal with such pain and perplexity within their subjective worlds, using suffering as a stepping-stone to wisdom and maturity. Approaches of this kind dominated Christian spirituality during the Middle Ages, when there was a widespread acceptance that suffering was simply a fact of life that did not require explanation; it did, however, need to be engaged with. Many writers of that age considered how someone could grow in wisdom and maturity through suffering, often using images of the crucified Christ as an imaginative gateway to reflection.[24]

Lewis's personal history helps us to see why this matters so much. To simplify what is a slightly complicated matter, we might say that *The Problem of Pain* represents a view from the Balcony, whereas *A Grief Observed* represents a view from the Road. Lewis's detached view of the suffering observed in other people proved inadequate to deal with his own personal experience on the Road, when it became a horrific reality in the life of someone he loved, and had a direct impact on his own feelings and life. Lewis did not reject the intellectual framework he developed in *The Problem of Pain*; rather, he realised its virtual uselessness *on the Road*. His earlier approach to suffering had been rationally adequate, but existentially deficient. It needed reformulation and contextualisation if it was to be of any use to those travelling through the valley of the shadow of death (Psalm 23). Lewis's correspondence of the summer of 1963 (by which time he knew he was dying) suggests that he achieved this transposition.

*   *   *

In the last few chapters, we have focused on the human search for meaning in life. Yet that quest for meaning also extends to the world in which we live. What is its significance? And what is our place within this world? In the next two chapters, we shall explore this broader quest for meaning, and the questions that it raises. We begin by looking at our attempts to make sense of our universe, and how this gives rise to the natural sciences.

8

Wondering about Nature:
The Imaginative Roots of Science

'I seem to have been only like a small boy playing on the
sea-shore, diverting myself in now and then finding a
smoother pebble or a prettier shell than the ordinary, whilst
the great ocean of truth lay all undiscovered before me.'[1]
Isaac Newton

The human experience of wonder is a gateway to understanding
our strange world. In a highly perceptive essay entitled 'The Illusion
of the Two Cultures', the American evolutionary anthropologist
Loren Eiseley argued that our artistic, humanistic and scientific
endeavours and achievements are all driven by the irresistible power
of the human imagination.[2] Science and art – and, I would add,
religion – are born of the same mind and are therefore inseparable.
Eiseley expressed concern that the contemporary academy, through
a relentless enforcement of disciplinary boundaries and its cult of
'professionalism', was draining a life-giving imaginative power
from the sciences. Such bureaucratic minds, he suggested,

> exhibit an almost instinctive hostility toward the mere attempt to
> wonder, or to ask what lies below that microcosmic world out of
> which emerge the particles that compose our bodies, and that
> now take on this wraithlike quality. Is there something here we
> fear to face, except when clothed in safely sterilised professional
> speech? Have we grown reluctant in the age of power to admit
> mystery and beauty into our thoughts?[3]

Eiseley worried that people had come to trust as real only those
objective truths revealed by science and the artificial world which

they themselves had constructed. The category of mystery was thus banished or declared to be redundant, meaning at most something that is not presently comprehended by science. But what if there is more to things than this?

Eiseley was critical of what we might call a detached, third-person approach to science, which treated nature as an impersonal object to be investigated, rather than as something to be respected, loved and admired. There had to be a way of restoring this sense of wonder, and reconnecting with a sense of being part of something greater. Human beings have the potential for stereoscopic vision, in that they can discern meaning within the world, and not simply figure out how things work. One of Eiseley's concerns was that technological progress was turning us into one-eyed creatures, capable only of seeing a reduced world, in which things are defined in terms of their function. Today's secular 'disruption between the creative aspect of art and that of science' represents an unnecessary and reversible fracture, dependent for its imaginative plausibility on 'the deliberate blunting of wonder'.[4]

In his best-known essay 'The Star Thrower',[5] Eiseley describes walking along a beach and encountering a boy throwing stranded starfish back into the ocean. Eiseley initially regarded this as pointless. As a scientist, he was perfectly aware that he should have no compassion for those starfish, as if they were reflective beings that cared whether they lived or died. Darwinian theory held that evolutionary progress required death, so why interfere with this natural process? Yet as a human being, he found himself realising that, as a scientist, he had missed something. 'It was as though', he writes, 'at some point the supernatural had touched hesitantly, for an instant, upon the natural.'[6] The next day, Eiseley was on the same beach, throwing stranded starfish back into the ocean. It was, he mused, both an act of renunciation of his scientific heritage, and an embrace of a greater vision of things.

Perhaps most importantly, Eiseley – as an evolutionary anthropologist – held that there was something about human nature that transcended its biological origins. The human sense of wonder held the key to the deepest questions of life.

> Man partakes of that ultimate wonder and creativeness. As we turn from the galaxies to the swarming cells of our own being, which toil for something, some entity beyond their grasp, let us remember man, the self-fabricator who came across an ice age to look into the mirrors and the magic of science. Surely he did not come to see himself or his wild visage only. He came because he is at heart a listener and searching for some transcendent realm beyond himself.

We long to understand what we find so beautiful and mysterious, suspecting that it might unlock access to a greater and deeper vision of reality. Many experience a sense of standing on the threshold of a partly glimpsed world, which seems to lie beyond the everyday reality that we inhabit, yet is somehow hinted at by what we observe around us and experience within us. Our experience of wonder somehow transcends a mere understanding of the world, seemingly enfolding us in some deeper relationship with the world we inhabit, and whatever might lie behind and beyond it.[7]

The roots of science ultimately lie more with this sense of wonder at the beauty and grandeur of our world than in a desire to understand it, still less to master it and redirect it to our own purposes. It is, of course, impossible to separate these three elements of the scientific enterprise. Inevitably, some will see the trajectory from wonder through to understanding through to manipulation and exploitation as a descent; others, however, see it as a necessary transition from merely intellectual reflection to the more serious business of ensuring human survival.

Yet that sense of wonder at nature remains fundamental to many, pointing to both the intellectual curiosity and sense of awe that many see as the primary motivation of the natural sciences. It is, however, all too easily lost through over-familiarity and a faulty understanding of scientific explanation. This calls for further exploration.

Disenchantment: A Loss of Wonder

Professional astronomers regularly tell me that they envy their amateur counterparts, who so often seem able to maintain their sense of awe at the beauty of the heavens, and see this as a driving force in sustaining their love of sky-watching. Most of them confess that they lost any such sense of awe years ago, eaten away by the acid of over-familiarity.

One of the most celebrated literary examples of the routinisation of wonder is found in Johanna Spyri's classic children's story *Heidi* (1880), which tells of a young girl's first encounters with the grandeur and beauty of the Swiss Alps, often in the company of Peter the goatherd, who knows the mountain pastures well. On one occasion, Heidi experiences a sense of wonder at the *Alpenglühen*, an alpine atmospheric condition which causes the mountain peaks to glow in a brilliant red colour at sunrise and sunset. Peter, however, has seen this phenomenon so many times before that it has become dull and routine. He dismisses her amazement with a shrug of his shoulders. 'It's always like that.'[8]

Yet this loss of a sense of wonder can arise for other reasons. The sociologist Max Weber used the term 'disenchantment' to refer to the processes by which the natural sciences eliminated any sense of mystery or wonder through creating 'a thoroughly rationalising view of the world'.[9] The natural phenomenon that is often singled out for special mention here is the rainbow – something which can evoke a sense of awe or wonder, but which is easily explained by optical physics.[10] The issue here, of course, is whether a scientific explanation of beauty detracts from the experience of beauty.

For some – such as the Romantic poet John Keats – science impoverishes our appreciation of the natural world, reducing everything to general principles:

Do not all charms fly
At the mere touch of cold philosophy?
There was an awful rainbow once in heaven:
We know her woof, her texture; she is given
In the dull catalogue of common things.
Philosophy will clip an Angel's wings,
Conquer all mysteries by rule and line.[11]

Keats took the view that any sense of awe in the presence of a rainbow was reduced to the cold and impersonal laws of physics by 'natural philosophy' – as the natural sciences were generally known in Keats's day.

Some hardnosed rationalists have dismissed Keats's concerns, seeing them as rambling nonsense that is typical of the uncomprehending anti-scientific attitude of poets in general.[12] Others, however, have realised that there is a real issue here. It is perfectly possible to give an objective scientific account of how a rainbow works, or of the atmospheric effects which often make sunsets seem so beautiful. Yet the fact remains that the subjective human response to nature matters to so many people. Scientific analysis is based on third-person perspectives, and thus takes an impersonal attitude which has difficulty in dealing with the intensely subjective human experience of beauty.

So why should a concern for the way we feel about nature be seen to be the *enemy* of science? It is important to maintain a balance between an objective scientific explanation of an aspect of the natural world and a subjective experience of its beauty or wonder. Knowing how a rainbow works does not prevent us from appreciating its beauty. Nor can it forbid us to ask what it might mean. Maybe that question cannot be answered with the cool and comprehensive clarity that many would expect. But it most certainly needs to be asked! And if there is indeed an answer to be found, it will change everything. For many, this subjective experience of wonder acts as a motivator for investigation and discovery – a point that we need to explore further.

One Outcome of Wonder: Natural Science

A sense of awe and wonder in the presence of natural beauty arises partly from our realisation of our limited grasp of the immensity of our world. So much seems tantalisingly beyond our grasp, yet we long to reach beyond our natural limits and horizons, and grasp what we sense lies beyond. Scientifically, this extension of our boundaries primarily takes the form of technological enhancement. The development of the telescope and microscope extended the natural reach of the human eye, and opened up new worlds for exploration and explanation.

The telescope is believed to have been invented in Italy around 1590.[13] It was initially seen as an optical toy which amazed people all over Europe in the early seventeenth century by its ability to enlarge distant objects, making them seem much nearer. Its potential value in the arms race of the age was quickly recognised by European navies and armies. Yet perhaps its most important applications were scientific. The telescope revealed that the heavens contained far more stars than expected, that there were mountain ranges on the moon, that there were moons orbiting Jupiter, and that there were spots on the sun.[14] The scientific revolution of the age may have demanded a technological enhancement of the natural human capacity to observe, yet it was ultimately based on the natural human capacity to wonder about our world, and our place within it.

To wonder about the world involves both the imagination and reason. The experience of wonder triggers a process of reflection, as we try to work out what 'big picture' of reality makes most sense of things. After a period dominated by an excessively rationalist account of scientific discovery, the philosophy of science has now come to recognise the importance of the imagination in the development of scientific theories.[15] A scientific theory is a 'way of seeing things' – as the Greek term *theoria* suggests, a way of beholding our world – which allows us to grasp how things hang together and relate to each other. Scientific theory is an act of imagination – not in the sense of making things up, but rather in the sense of enabling us to visualise patterns of relationships within the world that help us explain what we observe. As the American theoretical physicist Richard Feynman (1918–88) pointed out, the scientific enterprise relies on the human imagination, which often finds itself 'stretched to the utmost, not, as in fiction, to imagine things which are not really there, but just to comprehend those things which are there'.[16]

Werner Heisenberg argued that a good scientific theory would 'do justice to every new experience, to every accessible domain of the world'.[17] Yet while scientific theories will always do their best to develop a language that is suitable to the domain of reality that is under investigation, it is impossible to avoid the feeling that 'there are other phenomena that defy formulation in language'. There is a sense of mystery about our universe, in that every scientific advance simply opens up new questions, often calling into question the capacity of human minds and language to cope with the external reality that we call the universe. Yet scientific thinking 'always hovers over a bottomless depth', given the limits placed on human understanding. 'Every time when there is an understanding of a new reality, their sphere of validity appears to be pushed yet one more step into an impenetrable darkness that lies behind the ideas language is able to express.'[18]

Parallels between Science and Christian Theology

To anyone who is familiar with the Christian faith, there are obvious parallels with a scientific understanding of reality.[19] Many people find that their religious faith, like science itself, originates in and through a sense of wonder, which raises questions about their place within the world, and a deeper reality that might lie behind or beyond the world of experience. William James saw both science and religion in terms of a quest for an 'unseen order of some kind in which the riddles of the natural order may be found and explained'.[20] Once this 'bigger picture' is grasped by our imaginations, we are able to make far more sense of what we see and experience. It's a good point. Scientific accounts of natural processes are basically about the unfolding of an order that is already implicit in the nature of things, although often in an opaque or hidden way.

Yet there are limits to what we, as human beings, can discern of this reality on our own. In the case of science, we need to amplify what we can discern through our natural faculties – for example, extending the reach of the unaided eye through telescopes or microscopes. In the case of Christianity, we realise that our reason needs supplementation by revelation – the disclosure of a 'big picture' which we ourselves were unable to glimpse fully, yet which once given to us, makes so much sense of what we know and see.

When I was growing up, I took great pleasure in looking at the night sky through a little telescope that I had built, which allowed me to see the moons of the planet Jupiter, and track the movements of the planets against the background of the fixed stars. When I was twelve years old, I recall being puzzled by the movements of the planet Mars, which seemed to move backwards and forwards over a period of weeks.[21] I told my science teacher at school about what I had seen. He explained what is technically known as the 'retrograde motion' of Mars against the fixed stars by drawing some diagrams to show the relative motions of the earth and this planet. Since the earth moved faster than Mars around the sun, it caught up and then passed Mars's position relative to the sun. That was why it seemed to move in this strange way.

After about five minutes, the penny dropped. I could see what was happening. My teacher had given me a framework for understanding what I observed, and it made sense *in itself*, just as it made sense of *what I observed*. But I had not been able to work this out for myself. Someone had to show it to me. And once I was given this way of seeing things, which lay beyond my capacity to work out for myself, I found that it illuminated what I had observed. Revelation is about being given a 'big picture' of reality which we could not entirely work out for ourselves, but which turns out to be trustworthy, helping us make sense of what we experience.

Finally, both science and religion recognise the inability of human language to do justice to the complexity of the natural world. As we noted earlier (111), the great physicist Werner Heisenberg spoke of the 'bottomless depth' and 'impenetrable darkness' of the universe, and the human struggle to find a language adequate to engage with and represent this.[22] It is no criticism of science to suggest that it also fails to do justice to the subjective human response to the beauty and vastness of the universe. Those are the limiting conditions under which science operates.

Similarly, Christian theology recognises that it is utterly impossible to represent or describe God adequately using human language. Theology uses the term 'mystery' to refer to the vastness of God, which inevitably causes human images and words to falter, if not break down completely, as they try to depict God fully and faithfully. A mystery is not something nonsensical. Rather, it is something that exceeds reason's capacity to discern and describe – thus transcending, rather than contradicting, reason. To speak of some aspect of nature or God as a 'mystery' is not to attempt to shut down the reflective process, but to stimulate it, by opening the mind to intellectual vistas that are simply too deep and broad for our limited human vision.

The doctrine of the Trinity, which even many Christians find quite puzzling, is best seen as a principled attempt to prevent us from reducing our vision of God to what reason can manage, thinking we are making our faith more reasonable, when in fact we are distorting and diminishing God. We must expand our reason to cope with the reality of God, not limit God to what our minds can cope with! As the theologian Emil Brunner pointed out, the doctrine of the Trinity is a 'defensive doctrine',[23] designed to stop us impoverishing the Christian understanding of God by redefining God in the meagre terms of a thin rationalism, filtering out such notions as beauty and glory when they prove incapable of being accommodated by this emaciated way of thinking.

Yet perhaps one of the most interesting outcomes of the human sense of wonder at the beauty and vastness of the natural world is what is traditionally (although slightly misleadingly) known as 'natural theology'.[24] We shall consider this in what follows.

Another Outcome of Wonder: Natural Theology

Even the casual reader of the Old Testament will notice its interest in the natural world, and the connections it makes between the world of nature and God. The great closing divine speech in the book of Job superbly portrays the beauty, elegance and intricacy of a vast created order, which lies far beyond the capacity of any human mind to comprehend.[25] To contemplate the universe is to be forced to recognise the depths of the divine wisdom, and the inability of humanity to grasp the complexity of the world in which we live.

Perhaps the most famous Old Testament reference to the capacity of the natural world to point to God as its creator is found in Psalm 19:1:

> The heavens are telling the glory of God;
> and the firmament proclaims his handiwork.[26]

This verse inspired some of the most remarkable music of the eighteenth century, including Johann Sebastian Bach's Cantata *Die Himmel erzählen die Ehre Gottes* (1723), and Joseph Haydn's oratorio *The Creation* (1798). Its fundamental theme is that the heavens are able to declare, demonstrate or show the beauty and glory of God, their creator.[27]

So does the Psalm imply that we can *prove* God's existence by looking at the beauty of the created order? Surely not. Israel already knew about the existence of its God. The recognition that God's creation declares God's glory is not presented in this psalm as a mandate for deducing the existence and character of God by looking at the heavens. Psalm 19 affirms that a God whose existence and nature were already well known to Israel might be known in an aesthetically or imaginatively extended manner by reflection on what God has created – such as the beauty of the night sky.

The term 'natural theology' can be understood in at least six different ways, including the idea that the existence of God can be inferred from reflecting on the beauty or ordering of the natural world.[28] Yet the main theme of any Christian natural theology is that there is a link between the natural world and its creator. To grasp the beauty of what God has created helps us to appreciate the even greater beauty of God himself. It's a classic idea. One of my favourite ways of thinking about this is found in the writings of the medieval theologian Thomas Aquinas, who suggested that the beauty of nature was like a trickle of water in comparison with the fountain of beauty to be found in God.

So why is a natural theology so important to a Christian appreciation of the natural world? I've often reflected on this, particularly as a result of my debates with Richard Dawkins. Dawkins holds that belief in God impoverishes our appreciation of the beauty of the natural order; I think it enhances it.[29] Let me explain why.

Christian Faith and the Appreciation of Nature

Dawkins argues – rightly, in my view – that it is perfectly possible to have a sense of 'awe' or reverence for nature without being religious or believing in God. Unfortunately, he spoils this perfectly good point by his unevidenced assertion that belief in God actually *diminishes* this sense of awe, by offering an aesthetically deficient view of the universe which is 'puny, pathetic, and measly'[30] in comparison to the way the universe actually is. For Dawkins, the 'universe presented by organized religions is a poky little medieval universe, and extremely limited'. The logic of this assertion is frankly somewhat confused, and its factual basis is disappointingly slight. I certainly agree with Dawkins that some 'medieval' views of the universe were much more limited and restricted than modern conceptions. But this has nothing to do with religion, in that it was simply the scientific consensus of the age. If people believed in a 'poky' universe in the Middle Ages, it was because they trusted what their science textbooks told them was right.

Dawkins is certainly unreliable as a historian; I'm afraid that he also fails to grasp the difference that faith in God makes to the way we see and appreciate the natural world. Christianity gives us an imaginative and rational lens through which we can see nature more clearly, and appreciate it more fully. As far as I can see, there are three main elements to a Christian appreciation of the natural world.

First, we experience an immediate sense of delight or wonder at the beauty of nature – for example, a glorious sunset, a majestic snow-capped mountain range, or a verdant tropical landscape running down to a white sandy beach lapped gently by a warm turquoise sea. This experience rests neither upon a prior belief in God, nor upon the absence of such a belief. It arises independently of the mental map of the observer.

Second, one of the greatest achievements of science is the 'mathematisation' of nature – in other words, the representation of its relationships and structures using mathematical equations. As Eugene Wigner and others have pointed out, this is actually rather strange. Why should there be any correspondence between the natural world and mathematics? After all, mathematics is a human invention! Wigner himself declared that 'the miracle of the appropriateness of the language of mathematics to the formulation of the laws of physics is a wonderful gift which we neither understand nor deserve'.[31] When scientists set out to make sense of the complexities of our world, they use 'mathematics as their torch'. But why?

For Wigner, this amounted to a mystery that called for an explanation. Was there some 'ultimate truth', understood as 'a picture which is a consistent fusion into a single unit of the little pictures, formed on the various aspects of nature'? Christian theology, of course, provides such a 'picture', helping us to see why there is such a correspondence between mathematics and the structures of the created order.[32]

While this is an important point, there is something else that needs to be noted here. These mathematical representations of nature are themselves elegant and beautiful. Indeed, this is such a significant observation that the great physicist Paul Dirac suggested that beauty might be a criterion for truth. He illustrated this with reference to Einstein's theory of relativity.

> What makes the theory of relativity so acceptable to physicists in spite of its going against the principle of simplicity is its great mathematical beauty. This is a quality which cannot be defined, any more than beauty in art can be defined, but which people who study mathematics usually have no difficulty in appreciating. The theory of relativity introduced mathematical beauty to an unprecedented extent into the description of Nature.[33]

Now an appreciation of the beauty of a theoretical representation of nature – such as a mathematical equation – neither requires faith in God, nor the absence of such a faith. It is neutral. Yet the framework of meaning which is provided by the Christian faith helps us understand how the beauty of such equations arises within and corresponds to the beauty of God. The great astronomer Johann Kepler (1571–1630), for example, argued that, since geometry had its origins in the mind of God, it was only to be expected that the created order would conform to its patterns. Since geometry is 'part of the divine mind', should we be surprised that it 'provided God with the patterns for the creation of the world, and has been transferred to humanity with the image of God'?[34]

Yet perhaps it is the third point that is the most important. For the Christian, the natural world is embedded with signs pointing beyond itself to its creator. An appreciation of nature from a Christian perspective leads effortlessly and naturally to an appreciation of God as creator. This connection is, of course, entirely absent within an atheist framework of meaning. There is no transcendent domain to be signposted by the natural world.

For the Christian, however, the natural world bears the fingerprints of God. Our experience of delight evoked by the beauty of the world is only a hint of the greater delight evoked by the sight of God, who bestowed that beauty in the first place. To appreciate the beauty of nature is thus to anticipate an experience of the overwhelming loveliness of God. Where Dawkins suggests that Christians miss or impoverish the beauty of nature, the reverse is actually true. Christians are primed to discern that grandeur, knowing it will trigger off an exquisite longing to see God face to face.

It might be objected that this way of approaching nature is purely subjective. It has no objective reality in itself, open to rigorous scientific analysis. There is an obvious truth in this objection. Yet it misses a still greater truth: that the subjective world of experience matters profoundly to human beings, so that no philosophy or worldview that fails to engage with human subjectivity will ever secure the deep traction of commitment. As the philosopher Rudolf Carnap noted, a totally objective scientific account of things – such as Einstein's theory of relativity – 'cannot possibly satisfy our human needs'. What humans need to lead existentially satisfying and meaningful lives necessarily lies 'outside of the realm of science'.[35]

Yet even this concern about subjectivity needs to be challenged. *All* mental frameworks are 'subjective' – no matter how well they are evidenced. They exist within us, not beyond us, and frame our understanding and action. The same objective reality can be observed by two different people, and yet be experienced and 'seen' in quite different ways. We shall consider this point further when asking whether human beings can 'feel at home' in this universe – a classic example of a subjective perception which many people regard as deeply important.

9

At Home in the Universe?
Wondering about Our Place in the Cosmos

'When I consider the short duration of my life, swallowed
up in the eternity before and after, the little space which I
fill, and even can see, engulfed in the infinite immensity of
spaces of which I am ignorant, and which know me not, I
am frightened, and am astonished at being here rather than
there; for there is no reason why here rather than there, why
now rather than then. Who has put me here? By whose order
and direction have this place and time been allotted to me.'[1]
Blaise Pascal

Pascal's haunting passage identifies some of the deepest fears of
human beings – that we are insignificant elements in a vast and
meaningless universe, which has not the slightest interest in us. We
seem to have been inserted into the cosmos at a place and time
which were not of our choosing, and we will eventually depart
from it at a place and time which are unknown to us. Our 'world
line' began with our birth at (x^1, y^1, z^1, t^1), and will end with our
death at an indeterminate point in the future (x^2, y^2, z^2, t^2). So why
did we enter this cosmic process *then* and *there*? Or is it utterly
meaningless to ask the question *why*, when all that can be hoped
for is an answer to the rather different question *how*? And what
about that most troubling and perturbing question of all: *so what*?

It is not surprising that many have concluded that we have
found our way into this strange and puzzling world by accident.
The biologist Jacques Monod took this view in his book *Chance
and Necessity*. Human beings, he declared, are an accidental and
meaningless presence in an equally accidental and meaningless
universe. Man must 'realize that, like a gypsy, he lives on the

boundary of an alien world; a world that is deaf to his music, and as indifferent to his hopes as it is to his sufferings and crimes'.[2] We are not *meant* to be here. We find ourselves in a world which is neither of our making nor of our choosing. In fact, we are not meant to be *anywhere*. We just exist.

Nature is thus the prison in which we find ourselves trapped, the landscape through which we are passing on a journey that goes nowhere. The universe doesn't know we exist. It doesn't care. Like us, it just exists. And now we know that one day it will die – just as we too must die. It's a thought that is expressed lyrically in a poem by the twelfth-century Persian mathematician Omar Khayyám:

> And that inverted Bowl they call the Sky,
> Whereunder crawling coop'd we live and die,
> Lift not your hands to *It* for help – for It
> As impotently moves as you or I.[3]

So why do so many of us feel such a sense of disconnection with this universe? After all, every human being is made of stardust, the basic stuff of the universe. Elements such as carbon, nitrogen and oxygen were all forged in the cores of stars.[4] At the level of our fundamental building blocks, we are very much part of the universe. And yet so many feel alienated from it, sensing our solitude and isolation within the cosmic mechanism. We sense that there is – that there *has to be* – more to life than the universe itself can offer. We seem to be born with an innate capacity to wonder, intuiting that there is some greater vision of reality which might make sense of our presence and place in the universe. It's like the scientist confronted with a mass of observations, who knows that there has to be some way of joining up the dots which discloses a pattern that makes sense of things – a grand theory, which lies hauntingly beyond our reach.

Perhaps that experience of wondering might begin with the simple yet astonishing fact that there is indeed a universe, and that we are here to observe it – what Ludwig Wittgenstein termed an 'existential wonder' elicited by our imaginative awakening to the world around us.⁵ Michael Mayne captured this sense of exquisite sensitivity towards the existence of such a greater world while he was living in the Swiss Alps:

> My subject is wonder, and my starting point is so obvious it often escapes us. It is me, sitting at a table looking out on the world. It is the fact that I exist, that there is anything at all. It is the *givenness* that astonishes: the fact that the mountains, the larch tree, the gentian, the jay, *exist*, and that someone called *me* is here to observe them.⁶

Most of us have experienced some such sense of the strangeness of things – the sheer oddness of the fact that each of us is here, and able to see a reality beyond ourselves. G.K. Chesterton wrote movingly and perceptively of a suppressed, submerged or forgotten 'blaze or burst of astonishment at our own existence', lying at the back of our minds, but able to break through to inform and enrich our 'artistic and spiritual life'.⁷ How could people be 'made to realise the wonder and splendour of being alive', when their dulled imaginations could grasp little more than the quotidian?⁸

Yet that sense of wondering rarely terminates at this point. Having appreciated the singularity and surprise of our individual existence, we cannot help but wonder about the world in which we find ourselves, and our place within it. Is this universe our home – even if only for an astonishingly short time, in terms of the vast expanse of cosmic history? Or are we like sojourners, passing through it on our way to somewhere else? Do we really belong here?

Feeling at home in the universe is not a matter of cold logic or mathematical calculation; it is a deep-seated sense of belonging. Yet few actually feel at home in this universe. Some are merely resigned to their temporary occupation of cosmic space. Others feel deeply alienated, sensing that they do not belong in this world at all, but without any idea of where they might find their proper location. Let's reflect on these issues more fully.

On Existential Black Holes

The scientific term 'black hole' was introduced by the physicist John Wheeler to refer to an object or region within space-time that has such a strong gravitational field that nothing can escape from within it, including light.[9] It was not a new idea. In 1783, the English country parson and natural philosopher John Michell argued for the likely existence of what he called 'dark stars' that were sufficiently massive and compact to have such a strong gravitational field that light could not escape. Any light emitted from the surface of these stars would be dragged back by the star's gravitational field. The implications of this are far-reaching. For a start, you can't tell what is inside a black hole from the outside. Nor can you see them. Their existence has to be inferred indirectly from observations of their gravitational interactions with their surroundings. It's a classic example of the inference of the existence of something that cannot be seen from what can be seen.

Yet that same term was used in an *existential* sense in 1861 by the great French novelist Gustav Flaubert to refer to the 'black hole (*trou noir*)' that stood at the heart of human existence. In a letter to his friend Madame Roger des Genettes, Flaubert wrote of the existential void that stood at the heart of a godless world.

> The melancholy of antiquity seems to me more profound than that of the moderns, all of whom more or less imply that immortality lies beyond the black hole. But for the ancients, that black hole is infinity itself; their dreams take shape and pass away on a foundation of changeless ebony. No crying out, no convulsions – nothing but the steadiness of a pensive face. Just when the gods were no more, and the Christ had not yet come, there was a unique moment between Cicero and Marcus Aurelius when humanity was alone.[10]

What was Flaubert getting at? His major theme is that humanity, in the absence of the divine, is alone in the universe, steadfastly facing a meaningless void without any hope of immortality. Flaubert regarded such a resignation as noble. If happiness was to be found, it lay in the acceptance of human isolation within a faceless universe, realising that our beliefs in meaning are dreams which rise and fall within a 'black hole' of disinterested nothingness.

A similar view was taken by Martin Heidegger (1889–1976), a German philosopher who did much to encourage the emergence of existentialist philosophy, which recognised the important role of human subjectivity. For Heidegger, we have been 'thrown' into this world, and try to make sense of its riddles, including our own identity and purpose. We search for 'authenticity' or 'authentic existence', without entirely understanding what form this should take, yet possessing the freedom to choose what we believe to be the best option.[11] Heidegger himself seems to have believed that 'authentic existence' lay in the acceptance of human mortality, and in resisting any notion of an afterlife or transcendent domain.

But what happens if there is a God, and if this God chooses to enter and inhabit this world of ours – and in doing so, dignifies this 'black hole', lending it value and meaning, and saturating it with intimations of eternity and immortality? Flaubert hints at this possibility, perhaps realising that the coming of Christ changes everything. But it's a central theme of the Christian way of thinking about our relationship with this world, as we shall see later in this chapter. Before we consider this idea, though, let's think more about what it means to be at 'home in the universe'.

Home: Belonging Somewhere

The word 'home' is one of the richest and most emotive words in any language. It designates somewhere special – a place where we feel, above anything else, that we *belong*. It's a way of mapping space in our minds, highlighting one special place which is our spot. We come to feel a deep kinship, a sense of connectedness, with our home. Our heart belongs there, and we feel disconnected and alienated when we end up somewhere else. We find ourselves experiencing an existential ambivalence – a queasy unsettledness – which hints that this is not where we really belong.

It's part of being human to feel moored to a specific place, rich in associations and memories. Home, as they say, is where the heart is. Yet it proves hard to conceptualise what is distinct about a 'home', and what distinguishes it from a 'habitat'. Perhaps that's because 'home' is really a category of the human heart, rather than of the human mind. It's about the way we *feel* about a place – experiencing its existential magnetism, taking pleasure in its associated sense of peace and security, and feeling ill at ease when we are away. We are rooted animals, who often see 'home' as a repository for our personal historical pasts, a gathering place for memories that define who we are partly in terms of where we come from.[12]

For some people, the notion of home is problematic. What about the perpetual nomads of our global culture – the international players who move from one continent to another, never really setting down roots anywhere? Or what about immigrants, who leave behind their homelands and their family roots as they seek a better life, or freedom from oppression? They bring only their memories to their new situations. Yet many immigrants show a remarkable capacity to recreate the vestiges of their lost homeland in a different place, while adapting to a new language and society. Memories of their homeland subtly shape their understandings of their adopted country, helping them to cope with their sense of alienation and disconnection.

Where physics speaks of space and time, most of us find these notions abstract and disconnected from human concerns. We prefer to speak of place and history. The distinction is, of course, subjective. Yet subjectivity really matters to human beings, and it cannot be ignored. This point is emphasised by the noted Old Testament scholar Walter Brueggemann, who argued that, to make sense of the theological concerns of ancient Israel, a fundamental distinction had to be made between 'space' and 'place'. 'Place is space which has historical meanings, where some things have happened which are now remembered and which provide continuity and identity across generations. Place is space in which important words have been spoken which have established identity, defined vocation, and envisioned destiny.'[13]

Brueggemann's analysis of the history of Israel shifted discussion away from the abstract notions of space and time, and anchored it firmly to the experiential realities of human existence, which are much more naturally expressed using the categories of place and history. We merely exist in space and time; we live and act in places and in history. Not every space is seen as a place. The French anthropologist Marc Augé highlighted the role of 'non-places' in Western culture – such as airports, shopping malls, hotels, highways and subways.[14] We have no sense of 'belonging' here; we merely pass through them, on our way to somewhere else. Home is where we feel that we are rooted, and where we belong.

Places play a critically important role in human life, not least in that they function as anchor points for memory, identity and aspiration. Two observers might look at the same objective space, but see it – and *experience* it – in very different ways. Place is what a space is *for us*, shaped by memories that are often known only to ourselves. That's why some see home as the place to which they long to return, others as a place they never want to see again.

This is a highly subjective matter, in that there is no physical or biological difference between a space and a place. Yet the fact remains that human beings both make this distinction, and regard it as important. Two people might look at exactly the same landscape at the same time. One sees the place where she met and fell in love with her long-dead lover; the other sees a rather uninteresting example of the kind of deciduous woods you find at the base of a mountain in that region. Both observe the same phenomenon; they see something quite different.

So are we at home in the universe? Or do we really belong somewhere else? Throughout this section, I have emphasised the importance of subjective perceptions and intuitions – in other words, engaging our feelings and desires, not the cold calculus of logic. So are there feelings within us that might suggest that our home lies somewhere else? That this is not where we really belong? Do we possess some homing instinct for another world? Let's turn to consider this question.

A Homing Instinct? Intuitions of Another World

One of the reasons I read poetry is to be challenged to write better prose myself. It's not that I'm depressed by the generally wooden quality of my writing, and need to be challenged to improve it. It's that poets so often seem to find the right words to express something that eludes more scientific writers. A good example lies in Matthew Arnold's lines from 'The Buried Life' (1852), in which he speaks of an irrepressible and 'unspeakable desire' for knowledge, rising up within him.

> A longing to inquire
> Into the mystery of this heart which beats
> So wild, so deep in us – to know
> Whence our lives come and where they go.[15]

Arnold's poem is haunted by the fear of loss of identity, absence of meaning, and deprivation of purpose in life. Arnold believed that we too easily lose our way in life, often in spite of external successes and appearances. The language of the poem hints at the importance of emotions, intuitions and feelings in causing the poet to rethink and recalibrate his life, reaching for something that has been lost, forgotten or suppressed.

Arnold puts into words a feeling known to writers and poets since late classical antiquity: that something has been submerged or overlooked; that there is something that can be uncovered and retrieved, and by doing so, we open up a way of living and thinking that had been lost to us. We are prompted in this rediscovery by instincts deep within us, which somehow serve to first awaken and then redirect the course of our lives. It is as if – to use a classical biblical idea – God has implanted a longing for eternity in our hearts (Ecclesiastes 3:11), and however much we try to suppress or ignore it, this keeps bubbling up in our dreams and reflections.

It is a thought that was famously explored (though in subtly different ways) by Augustine of Hippo in his *Confessions*, and by Blaise Pascal in his *Pensées*. Yet perhaps the most familiar exploration of this theme is that of C.S. Lewis, who was profoundly influenced by his deep intuition that there was more to life than his early 'glib and shallow rationalism' suggested or permitted. Lewis found that his reason was like a prison guard, trying to enforce the materialist 'spirit of the age'; his imagination screamed in protest, insisting that there was more to reality than this grim dogma allowed.

One of Lewis's most imaginative and engaging explorations of this theme is found in his sermon 'The Weight of Glory', preached at the Oxford University Church of St Mary the Virgin on 6 June 1941.[16] Lewis spoke of the haunting fragrance of another world, which spiced the stale air of our everyday experience, causing us to long to trace this scent to its source. In particular, Lewis highlighted the apologetic significance of a sense of longing for something – or someone – indefinable yet seemingly irresistible. So often such exquisite longings broke the hearts of those who experienced them, who mistakenly confused a sign with what it signified. This desire was a pointer to the one who alone could satisfy this seemingly unquenchable thirst, and to our transcendent goal, where such desires would find their final fulfilment.

Lewis summed this up in a landmark statement in *Mere Christianity*, perhaps his best-known work: 'If I find in myself a desire which no experience in this world can satisfy, the most probable explanation is that I was made for another world.'[17] To some, this seems a hasty, even premature, conclusion. Yet Lewis's point is that there are only two alternatives: to dismiss such longings as meaningless, or to abandon any search for their goal out of sheer exhaustion or frustration. We must never despise these longings on the one hand, or fail to appreciate that they are 'a kind of copy, or echo, or mirage' of what really matters. They are not the goal itself; they are the stimulus that made us look for and long for that goal in the first place.

Some will be worried that Lewis places such an emphasis on human subjectivity. Two responses can be made. First, elsewhere in his writings, Lewis develops much more objective modes of reasoning, which emphasise the explanatory capaciousness of the Christian faith. Yet perhaps the second response is more important. For Lewis, the subjectivity of the individual matters profoundly. The way we feel about ourselves, our world and God has a deep impact on our lives. Christianity effects an affective transformation of our emotional worlds, not simply a realignment of our thinking.

We simply cannot ignore the question of what it *feels* like to exist in the world. Some theologians are profoundly nervous of questions such as this, which seem to reduce the Christian faith to subjective emotions and longings. Now there are some reasonable concerns here, not least the risk of collapsing into some imaginary world that we have invented purely in order to satisfy our longing for meaning, or our feelings of inadequacy.[18] Yet there is a danger that this legitimate concern for a truth that is independent of us blinds us to the experiential, emotional and relational aspects of faith.

The Christian 'big picture' allows us to see the world in a new way. No longer is it a faceless void; it bears the imprint of God. Psalm 8 offers perhaps one of the most eloquent reflections on this theme. We may feel overwhelmed by the immensity of this universe, with its vast starry skies. Yet God has placed us within this universe, where we are meant to be. It is stamped and studded with signs of God's presence and glory; and we, who bear God's image, can discern those signs, and, illuminated and empowered by grace, may reach out and embrace the God who created us and our world.

Psalm 23 speaks of a journeying God who accompanies and guides us as we travel through this world. This insight is intensified and solidified by the distinctively Christian idea of the incarnation, which affirms that God entered into our world of history in Jesus of Nazareth. This intellectual framework allows us to make sense of our existence in the universe. Yet it does more than help us understand ourselves; it also enables us to live fully and authentically on the Road. Let's explore this point in more detail using a biblical image which has come to play a critical role in Christian spirituality: the idea of being exiled in this world.

The Exile: A Framework for Reflection

Why do so many feel that we don't really belong in this world? Such a sense of displacement or disconnection underlies the movement known as 'Gnosticism', which flourished in late classical antiquity.[19] Central to many forms of Gnosticism was the belief that we do not really belong here. We are meant to be somewhere else. We are trapped in this universe, and struggle to find our way out of it, and return to the realm where we our true destiny lies. Human beings are 'gold in the mud', spiritual beings who are imprisoned in material bodies and a corrupt world. Our ultimate goal is to escape from these prisons, and make our way to our true homeland.

It's a feeling that was well known to G.K. Chesterton, who had his own explanation of its origins and significance. We experience a sense of homesickness for our real native land, as we slowly realise that this world is not where our true destiny lies. 'We have come to the wrong star . . . That is what makes life at once so splendid and so strange. The true happiness is that we *don't* fit. We come from somewhere else.'[20]

While this experience is not specific to Christians, the Christian tradition brings a particular framework of interpretation to bear on it. The New Testament letters – often using the imagery of citizenship to make the point – emphasise that this world is a place of transit, rather than our permanent habitation. We have no permanent citizenship in this world, in that our citizenship is in heaven (Philippians 3:20). 'Here we have no lasting city, but we are looking for the city that is to come' (Hebrews 13:14).

Christian theologians developed other organising images to help illuminate this idea. One of the most powerful derives from first-century Roman imperial culture. Roman citizens would regularly serve the imperial administration abroad – for example, in the colonies of North Africa or Asia Minor – eagerly anticipating their return home on completion of their service. Rome was their *patria*, their 'homeland' – the place where they really belonged.[21] Early Christian writers used this framework to illuminate the dynamics of living on earth, while hoping for heaven. Cyprian of Carthage, a Roman citizen who converted to Christianity and served as bishop in the great colonial city of Carthage in North Africa in the third century, expressed this way of seeing things in a pithy maxim: 'paradise is our native land (*patria*)'.[22] Yet the framework that many regard as best adapted to express and capture the experiential aspects of the Christian way of thinking about our place in this universe is that of exile. A landmark episode in the history of Israel was developed by Christian spiritual writers into an imaginative template, allowing Christians both to visualise and to evaluate their position within the world of space and time. Although it offers a Balcony view of life, it comes into its own as a way of thinking about faith on the Road. The historical episode is what is generally known as the 'Babylonian Deportation' or 'Babylonian Exile', a complex event which extended throughout much of the sixth century BC.[23]

The exile began with the destruction of the city of Jerusalem and its Temple and the ending of the Davidic monarchy in 586 BC. Following a failed rebellion by the kingdom of Judah against the Babylonian Empire, Nebuchadnezzar besieged the city of Jerusalem, and deported most of its inhabitants over the period 597–581 to Babylon. Archaeological evidence indicates that most of the population of rural Judah remained in place; those who were deported were mostly from Jerusalem itself. They would remain in exile until the fall of Babylon to Persian king Cyrus the Great in 539 BC. Although some Jews remained in Babylon, most appear to have made their way back to Jerusalem in the following years, rebuilding the fallen city walls and its Temple.

The narrative of this historical event is important in its own right. Yet it gave rise to an imaginative template, a way of looking at the world, which was widely adopted and developed within the Christian spiritual tradition, especially during the Middle Ages. Christians, it is argued, should locate themselves within an imaginative framework in which they are the inhabitants of Jerusalem who have been exiled to Babylon. They do not belong there, and long to return home. During their enforced absence, they console themselves with singing the songs of their lost homeland, keeping its memory alive, while they wait for the day when they can go home. In the meantime, they live in hope in Babylon, knowing their true identity and destiny. The great medieval theologian Peter Abelard captured this in one of his most famous hymns:

> Now, in the meantime, with hearts raised on high,
> we for that country must yearn and must sigh,
> seeking Jerusalem, dear native land,
> through our long exile on Babylon's strand.[24]

This does not encourage any disengagement or disconnection from the world; it simply affirms that we really belong elsewhere. Yet this world is our place of transit in which we begin to garner hints that something lies beyond it. Like the prisoners in Plato's dark underground cave, we sense there must be more than a world of shadows. The theologian Josef Pieper noted how our natural hopes and longings were grounded in something beyond us, yet not alien to us. 'All our natural hopes tend towards fulfilments that are like vague mirrorings and foreshadowings of, like unconscious preparations for, eternal life.'[25] Such enigmatic foreshadowings disclose both the penultimacy of our own world, and the distant lure of another, which is our true home. For Abelard, this world is God's world, and is to be valued, appreciated and enjoyed. Yet it is studded with clues that it is not our real home; that there is a still better world beyond its frontiers; and that we may hope one day to enter and inhabit this better place.

Although the image of 'exile' is grounded in the theological framework of the Christian faith, it has found wider cultural resonance, not least because it offers a framework for naming and understanding the feeling of not being at home in the world. For the literary critic Edward Said, the notion of exile speaks of an 'unhealable rift' imposed between an individual and a 'native place', between a person and their 'true home',[26] evoking a sense of loss and alienation. Simone Weil saw the notion of being 'rooted' as a core 'need of the soul'.[27] Yet the concept of 'rootedness' does not entail belonging *here*; rather, it is about belonging *somewhere*, which makes existing *here* bearable.

Thus far, we have reflected mainly on our present situation, above all our quest for meaning in this universe. So what of the future? What hope might we have for ourselves and our world? In the final section of this work, we shall consider some of the deep and troubling issues that human beings face, and how we might engage with these.

Part Three

Wondering about Our Future

IO

What's Wrong with Us?
Why We Need the Idea of Sin

'If only it were all so simple! If only there were evil
people somewhere insidiously committing evil deeds,
and it were necessary only to separate them from the rest
of us and destroy them. But the line dividing good and
evil cuts through the heart of every human being. And
who is willing to destroy a piece of his own heart?'[1]
Aleksandr Solzhenitsyn

What is the future of humanity? Nobody knows. For a start, we
might suffer the same fate that is thought to have wiped out the
dinosaurs – an 'extinction event' caused by collision with a mete-
orite or comet.[2] These events are certainly unpredictable; with
due respect to the imaginations of Hollywood movie-makers,
they are also probably unavoidable. Yet most worryingly of all,
human beings have developed technologies that now put them in
a unique position. We, alone among earth's species, are able to
bring about our own mass destruction – perhaps through the
reckless use of nuclear weapons or designer pathogens. Never
before has a species emerged which might bring about its own
extinction event.

But why, some will ask, would humanity do something so
utterly stupid and perverse as to bring about its own destruction? If we were the idealised rational calculating machines of
the 'Age of Reason', we would never dream of doing something
so bizarre. Yet as we reflect on our future, we need to ask an
awkward question. Are we really that clever? Or is there something wrong with us, that allows or even impels us to do some
very unwise things?

We need to understand ourselves – who we are, what we are meant to be, and why we so often get things so badly wrong. I'm not a great fan of the English essayist William Hazlitt (1778–1830). Yet some of his aphorisms strike home powerfully, their crystalline prose softening the blow of the hard truths they contain. That's certainly true of his comment on human nature. 'Man is the only animal that laughs and weeps; for he is the only animal that is struck with the difference between what things are and what they ought to be.'[3] Both Hesiod and Plato spoke of a lost 'Golden Age' of harmony, stability and prosperity, whose memory tantalisingly haunted the present, intimating that this was not the way things were meant to be. The rich and complex Christian worldview frames human history in terms of fleeting memories of a lost paradise and the hope of its future restoration. Something is wrong with us. But what? Is there something that can be done about it? And how can we make sense of the vast gulf between what things are and what we feel they ought to be?

Human history is littered with bright hopes and dismal failures; with technological inventions that some unwisely hoped might end war and suffering, yet which seem to end up being used to promote them. How are we to make sense of this enigma? What 'big picture' of human nature helps us make sense of the perplexing and generally discouraging patterns of history? Or maybe even to change them for the better? What story can be told that gives a coherent account of our dilemma and its possible resolutions?

Framing the Human Dilemma

During their Grand Tour of Europe in 1765–6, the great English man of letters Samuel Johnson and his biographer James Boswell met the Italian playwright Giuseppe Baretti, best known for his unevidenced attribution of the words *eppur si muove* ('nevertheless it moves') to Galileo, more than a century after his death. During their conversations, Baretti rubbished the idea of the fundamental goodness of humanity with a well-judged one-liner: 'I hate mankind, for I think myself one of the best of them, and I know how bad I am.'[4]

It was a witty remark, spoken largely in jest, and greatly appreciated by Boswell – perhaps because it offered a witty counterfoil to the prevailing platitudes of the 'Age of Reason', which took an elevated view of human rational and moral capacity. It was an age of boundless optimism, driven by a sense of the irresistible advance of rational creatures to ever-greater heights and achievements. This view was given a new injection of energy in the nineteenth century, as Ludwig Feuerbach and Karl Marx offered critiques of religion which treated religious beliefs as human productions, lacking any counterparts in the real world. God was not so much dethroned as declared to be a pointless fiction; human beings were thus unaccountable to any higher authority. Wars arose because of a lingering heritage of religious bigotry. Neutralise religion, and the world would become a peaceful place.

Yet it has all gone wrong. Perhaps Western thinkers were lulled into a false sense of security by the virtual absence of global conflict during the period 1815–1914, and prematurely concluded that human violence and barbarity were a thing of the past, abolished by the rise of science and rational thinking. Yet as the global explosion of war and destruction in the twentieth century made uncomfortably clear, the volcano of human violence had merely become temporarily dormant, not permanently extinct.[5] Further eruptions are to be expected.

As we saw earlier, Giovanni Pico della Mirandola declared that humanity had the capacity to be angels or beasts. So what if there is something within us that predisposes us towards the bestial, rather than the angelic? The brutality of the last century inclines us to suspect that this might well be the case, helping us to reconnect with older and wiser accounts of the human predicament. Perhaps we have come to the belated recognition that the Enlightenment hope of a more peaceful and humane world – laudable in itself – now depends upon human beings adopting a more realistic and resilient view of human nature than those idealised fictions of the Enlightenment itself.

In his chillingly accurate depiction of human violence during the twentieth century, Jonathan Glover speaks of the need to 'replace the thin, mechanical psychology of the Enlightenment with something more complex, something closer to reality'.[6] We need to 'look hard and clearly at some monsters inside us', if we are ever to cage and tame them. That's why we need a solid diagnosis of human nature, rather than delusional and aspirational promissory notes, which soon prove their lack of credibility and utility.

We want to think that we are good, and conveniently ignore evidence that subverts that belief. Human beings are very good at refusing to admit that they are wrong. They shift goalposts, rewrite narratives and redefine terms in a splendid but futile attempt to evade admissions that they were wrong.[7] A defeat is redefined as a victory. Sure, it's totally irrational. But it's also very human. That's just the way we are. And unfortunately, human beings like to think that we're fundamentally good, and lock ourselves into interpreting everything in ways consistent with this, filtering out anything that challenges this dogma. Yet everyone knows it's not right. So let's try and work out what is wrong with us, and what we might do about it.

False Solutions to the Human Dilemma

The dogma of the intrinsic goodness of humanity is a patently false notion which can only be sustained in five highly creative, but intellectually implausible, ways.

1. By denying there is a problem at all. Any suggestion that there is a flaw or defect in humanity is dismissed as irrational gibberish. So we sanitise our language to persuade ourselves that our flaws and failings are actually virtues and strengths. We do not fail; we just achieve a limited degree of success. We

do not tell lies; we simply offer an interpretation of things which others find disagreeable. The problem always rests with someone else, not with me. It is a classic way of hiding from truth (and from ourselves) through the recalibration of words, designed to sustain an illusion rather than help us come to terms with reality.

2. By ignoring history, or presenting its narrative in such a selective way that disconfirming evidence is simply airbrushed out of the picture, in much the same way as disgraced Soviet leaders of the 1920s used to be replaced with potted plants in doctored photographs of this bizarre age.

3. By equating 'being good' with 'being human'. This sleight of the mind means that human beings are good *by definition*, so that no refutation or contradiction of this unevidenced assertion is possible. This outrageous category violation simply locks us into a morally complacent self-congratulatory world, in which we are applauded for what we are, rather than challenged to become what we *ought* to be, or *might* become.

4. By declaring that 'good' and 'evil' are simply social constructions without any basis in an objective reality. Whether we are deemed to be good or bad thus depends on the prejudices and precommitments of those who are judging, not the qualities of those who are being judged. As a result, 'good' and 'evil' are seen as little more than the crystallised prejudices of those who seek to direct opinion, not valid statements about the real world with any diagnostic capacity.

5. By declaring that humanity can be separated into two categories: good and evil people. The latter are responsible for the evils of this world, whereas the former embody the fundamental goodness of humanity which is so conspicuously absent from the latter. This neat binary solution contains within itself an ethical imperative: since evil can be located within a specific group of human beings, any viable attempt to safeguard human goodness entails that those who are evil must be shunned, isolated or destroyed.

History and present experience alike point to human beings having a capacity for doing some good, even inspiring, things, matched by what appears to be an equally great capacity for messing things up and getting things wrong, which often – but happily not invariably – leads to evil and the infliction of pain. Augustine of Hippo, one of the early church's most important thinkers, argued that things that were good in themselves could nonetheless become the carriers of moral corruption, countering a simplistic equation of 'evil' with 'that which can transmit contamination and decay'.[8] We need a richer vocabulary to cope with the sheer complexity of human engagement with the world, and its morally variegated outcomes.

The awkward truth about human beings is that they are perfectly capable of taking good things and doing some thoroughly nasty things with them. Science and religion can both go seriously wrong. It's easy for ideologues on every side to seize on their failures, unburdened by complicating facts, and depict these as disclosing universal truths. Yes, both science and religion can go wrong – *badly* wrong. But that doesn't mean that either of them is bad. It just shows that they are both thoroughly human undertakings.

Perhaps the problem goes even deeper than this. Many seem reluctant to recognise that human beings can *willingly* do evil. The charming but utterly naïve belief that we 'needs must love the highest when we see it' (Tennyson) fails to do justice to the complexity and mixed motivations of human beings. We are all capable of doing – and being – good and evil. There may be some variation in the extent of this capacity, yet most of us are honest enough to recognise this tension within us, which pulls us in different directions.

Our problem is that we seem to lack a vocabulary adequate to describe this complexity within human nature. This is the disturbing enigma that any defensible and viable account of human nature must acknowledge and engage with. Evil is not located in the 'other' – in someone or something else. It is a living presence, whether dormant or active, within each of us. Aleksandr Solzhenitsyn's experiences in the Soviet labour camps led him to realise that the 'line dividing good and evil cuts through the heart of every human being'. Evil is not something that can be conveniently located in the 'other' – someone or something else that contaminates us. Its roots lie deep within us. We like to portray ourselves as the innocent victims of evil, preferring to overlook the awkward truth that we are perfectly capable of actualising our innate evil tendencies in certain contexts.

Let's explore this question by looking at how human beings have done some worrying things with science.

The Corruption of Scientific Innocence

Human beings are very good at messing things up. Just about everything they touch has the capacity to crumble to dust. Science is one of the greatest forces within today's world. It's capable of doing some great things. But it's opened the door to some horrors. I remember a debate at Cambridge University some years ago on the place of science in today's world. One speaker declared that science had saved the human race. It had given us penicillin and new medical techniques for saving and extending life. He was immediately challenged by angry audience members to tell the full story, not just the bits that fitted his Panglossian view of science. What about atom bombs and other weapons of mass destruction? Weren't those also produced by scientists? It was an uncomfortable moment. It was as if a curtain had been drawn aside, allowing us to see a dark secret that everyone preferred to keep hidden.

The moral ambiguity of science arises partly because science is ethically blind, and partly because it is undertaken and applied by human beings, with morally conflicting principles. My own reflections on this theme go back to the time when I was studying chemistry at Oxford University in the early 1970s. At that time, I took a strongly optimistic and positive attitude to natural sciences, seeing them as an intellectual undertaking which was wonderful in its own right, and which also opened the way to the extension of human life and the elimination of disease and poverty. And I was right about that – but only in part.

One of the core elements of Oxford University's chemistry course at this time was synthetic routes for complex organic compounds. The leading authority on this subject was Louis Frederick Fieser (1899–1977), whose classic series *Reagents for Organic Synthesis* gave me lots of ideas for my own work. Fieser served as Sheldon Emery Professor of Organic Chemistry at Harvard University, and had an outstanding international reputation in his field. He and his wife Mary Peters Fieser (1909–97) pioneered the artificial synthesis of a series of important naturally occurring compounds, including the steroid cortisone and Vitamin K, which was needed for blood coagulation.[9] Their brilliant synthetic procedures made medically important chemicals much cheaper and more widely available, with highly beneficial outcomes for patient care. Fieser and Fieser's work became increasingly important during the Second World War, as supplies of quinine (a naturally occurring compound used to treat malaria) began to run low.

That was all I knew about Louis Fieser when I studied as an Oxford science undergraduate in the early 1970s. It was only later, as a graduate researcher in Oxford University's Department of Biochemistry, that I learned that Fieser had also been involved in another project during the Second World War. The US Army urgently needed a chemical weapon suitable for eliminating troop concentrations in jungles and destroying population centres in the Pacific war theatre. Fieser and his team of chemists at Harvard won the contract to develop this weapon of mass destruction. They invented 'napalm', a petroleum gel that stuck to buildings and human bodies. Once ignited, it could not be removed or extinguished. Tests by the US Army's Chemical Warfare Service confirmed it was an ideal weapon for firebombing Japanese cities. Napalm proved to be a phenomenal military success, killing more

Japanese civilians than the two atomic bomb blasts of 1945 combined.[10]

When I first learned about this, I found myself deeply conflicted about Fieser. How could someone who had pioneered ways of synthesising chemicals that would extend human life and enhance its quality also create a chemical that was specifically designed to end life on an industrial scale, in a horrifically inhumane manner? The impact of napalm on Japanese soldiers and civilians was psychological, not merely physical, evoking a horror of being burned alive in a hideously painful manner.[11] My perception of Fieser changed, becoming darker and more unsettled, partly reflecting my difficulty in holding these aspects of his professional career together in my mind, and partly because this triggered off a more troubling question. What if this moral ambiguity in a leading scientist was also embedded within science itself?

It was not easy for me to accommodate this radical dissonance in my thinking about science, although I later realised that many physicists experienced similar anxieties about the military use of their science to develop nuclear weapons. Gradually, I came to the conclusion that science was not good; it was neutral. It was something capable of being used for good or for evil. It all depended on the people who were using it, and their motivations. *Science* is morally ambivalent because *human beings* are morally ambivalent.

So how can we make sense of this? And what can we do about it? For a start, we can face up to the problem.

Facing Up to the Human Problem

Rowan Williams once remarked that there is 'inbuilt into human beings a sort of dangerous taste for unreality'.[12] We prefer make-believe worlds in which everything is sweetness and light – apart, of course, from the worlds of those we dismiss as out-groups, who are invariably deluded, evil and stupid, locked into primitive affective states that ought to be relegated to where they belong

– an earlier stage in hominid evolution. We devise worldviews that are designed to reinforce our own importance, and denigrate those of people we despise. We end up becoming willing prisoners of our self-serving myths that do nothing to advance knowledge or understanding.

The philosopher John Gray's *Straw Dogs* (2002) was an iconoclastic book, caustically debunking the pretensions of the kind of bland humanist philosophy you find circulating with the Zinfandel at metropolitan dinner parties. For Gray, 'humans cannot live without illusion' – such as a blind faith in progress, or the goodness of human nature.[13] Humanists may like to delude themselves that they have a rational view of the world, yet their core belief in moral progress is a 'superstition', which is arguably further from the truth about the human animal than any of the world's religions. Progress in science and technology is subservient to selfish and corrupting human agendas, and does not inevitably lead to social and political progress. 'Without the railways, telegraph and poison gas, there could have been no Holocaust.'[14]

In place of such illusions, Gray offered a ruthlessly Darwinian account of human nature, which discards such cherished notions as that of progress as a pre-Darwinian myth. 'Humans can no more be masters of their destiny than any other animal.'[15] We have to realise that grandiose social experiments to remake our world are destined to fail, precisely because they are developed by human beings, and depend on delusion-prone humans to trust and implement them. We must learn to live without the consolation of religion, of scientific explanation, or any dream of a perfect society. After all, human beings didn't evolve so that they could find truth or meaning in life. They evolved to reproduce.

Where Richard Dawkins dismissed religion as *the* defining human delusion, Gray sees religion as one such delusion among many others, including those unacknowledged delusions entertained by *bien pensants* such as Dawkins himself, who somehow believe that criticising religion establishes the rationality of their own ideas. Such people think that naming a delusion entertained by other (lesser) people somehow validates their own beliefs, when it merely blinds them to their tenuous evidential basis.

Many humanists will dismiss Gray's view that 'humans are weapon-making animals with an unquenchable fondness for killing'.[16] Yet this is an evidenced claim that cannot be dismissed because we don't like it. It may be an overstatement. But I would rather have an overstatement of an uncomfortable truth than a blind refusal to engage with the substantial body of historical evidence that points to this inconvenient insight.[17] Gray's suggestion that 'death camps are as modern as laser surgery' may need some nuancing, but there is sufficient truth in his argument to challenge the delusions of even the most incorrigible rationalist. Gray has perhaps taken to heart R.G. Collingwood's famous remark of 1939 rather more than some of his more idealistic fellow philosophers: 'The chief business of twentieth-century philosophy is to reckon with twentieth-century history.'[18]

The evidential basis for abandoning the optimism of some leading Enlightenment thinkers about human nature is overwhelming, and I can see no good reason for maintaining this myth other than a forlorn dogmatic commitment to an outdated ideological orthodoxy. In any case, we now seem to have finally realised that many textbook accounts of the Enlightenment are actually 'thinly veiled ideological manifestos or pale reflections of current trends'.[19] In other words, we tend to project contemporary progressive intellectual and social outlooks back onto the 'Age of Reason', reconstructing it in our own image. Happily, intellectual historians have shown that the Enlightenment was a much more diverse movement than earlier generations realised, and it is not difficult to find writers within the movement – or on its fringes – who developed what can now be seen to be more realistic accounts of human nature, including the limits of human reason. Alexander Pope's *Essay on Man*, which we noted earlier (28–29), offers some perceptive and cutting reflections on human nature that allow us to begin to make sense of the horrors of modern genocide.

Yet Gray's point – however overstated – about genocide being 'as human as art or prayer' raises the question of what vocabulary we can use to engage with this and other human deficits, whether we see these as originating from within our evolutionary history or from somewhere else. In the following section, we shall revisit the Christian language of sin, which affirms and attempts to explain humanity's rationally inexplicable yet seemingly inescapable tendency to debase and destroy its own best achievements.

Why We Need to Talk about Sin Again

My own transition from atheism to Christianity was catalysed by a growing realisation of the importance of 'big pictures'. As a scientist, I took the view that the ability of a theory to account for observation and experience was an indicator of its truth. I came to the view that the Christian 'big picture' fitted in the world much more persuasively than its atheist counterpart.[20] I do not expect everyone to share this belief, but it is important to understand what this influential and immensely generative way of seeing things is all about, and why so many find it helpful in making sense of the enigmas and seeming contradictions of human nature.

The enlightened philosophers of eighteenth-century France dismissed the notion of sin, regarding it as insulting to human beings. It suggested that they were flawed and fallible, prone to selfishness and violence. Critics of the Enlightenment made the obvious counter-argument that the irrationality and violence of the 'Reign of Terror' which followed the French Revolution, seemed to confirm precisely these tendencies, and called out for them to be recognised and expressed properly. However archaic its language may seem, the vocabulary of sin engages with the fundamental ambiguity of human nature, and challenges naïve utopian visions of the human future.

Why is it that every human institution seems to subvert its own goals? The Christian Church, considered as an institution, regularly falls victim to social forces and pressures – such as the need to accumulate resources in order to continue its mission – which end up compromising its core values. And it's not on its own. Countless institutions, religious and secular, find themselves failing and collapsing through human flaws. For example, in the first decade of the twenty-first century, the United Nations sent 'peacekeeping' troops to protect vulnerable communities in Africa. What happened? These troops ended up raping and abusing local women, giving rise to a new social problem – 'peacekeeper babies'. United Nations Secretary-General Ban Ki-moon described this outrageous sexual abuse by peacekeepers as 'a cancer in our system'.[21]

Yet as the theologian Reinhold Niebuhr so prophetically remarked, the flaws in human institutions ultimately arise from corresponding flaws in human nature itself.[22] The 'cancer' lay not only in the 'system', but in its constituent human beings. This was one of the reasons why Niebuhr came to develop such a respect for the American constitutional system, with its checks and balances, and its genius for recognising the inevitable conflicts that would arise within society, demanding workable means of preventing the excessive localisation of power.

We cannot allow the delusion of a fundamental and incorrigible human goodness to shape such important issues as social policies or ethical thinking. The empirical realities of life demand that we realise that there is something wrong with us, and force us both to confront this awkward truth, and figure out how to minimise its impact. The Christian tradition offers what it regards as a totally realistic account of human nature, which is dismissed as 'pessimistic' only by those who close their eyes to what is happening in the world.

Christianity offers an evaluation of human nature. This can be framed as a forensic judgement that we are guilty or negligent, having failed to pursue goodness and combat evil. Perhaps we all need our delusions of moral perfection to be challenged from time to time in this way. Yet this evaluation of our condition can also be framed as a *medical* judgement – a diagnosis which tells us what is wrong with us, what might be done to cure us, or what could help us to manage a chronic and persistent condition that prevents us from being the people that we are meant to be.

The Christian understanding of human nature has two key points of focus: the notion of humanity bearing the 'image of God', and the concept of sin. Although there is some latitude of interpretation concerning both these ideas, there is a generous consensus within the Christian tradition about what these core notions are trying to express. The idea of humanity bearing the 'image of God' speaks of some inbuilt drive within human nature, which perhaps could be conceptualised as a 'homing instinct' for God. Just as a compass needle is drawn towards the magnetic pole, so the human imagination is drawn, as much by intuition as by reason, towards its origin and goal in God.

Perhaps more importantly, the notion of the 'image of God' expresses the notion that human beings are drawn upwards, avoiding collapsing into the material order from which they emerged. We emerge from dust; we shall return to dust; but we are more than dust. The human urge for self-transcendence can be seen in the light of this theological framework as a hidden or disguised yearning for God, which calls out for interpretation.

The term 'sin' is used in a theological sense to designate a flaw within human nature which prevents us from achieving our true goals. It is not a moral or existential concept; it is essentially a theological notion, with moral and existential *outworkings*. The term sin can be used to designate both individual actions which represent a failure to achieve our true goal, and the underlying human state that gives rise to those individual acts of sin. Writers such as Augustine suggested that human nature could be seen as damaged, wounded and broken. It needed healing and restoration if we are to achieve our true aspirations and goals. We are trapped – by our evolutionary past, by our personal weaknesses, and by the seductive whisper of delusions that have become the received wisdom of our day.

If the word 'sin' didn't exist, we would need to invent it. But it's already there, precisely because earlier generations recognised the problem, and knew that the first step in dealing with this problem was to name it. In and through its concept of sin, Christian theology gives us a critical lens through which to view the complex motivations and mixed agendas of human beings. We bear God's image, yet we are sinful. We are capable of good, just as we are capable of evil. If the 'image of God' affirms our need to reach upwards – to grasp and be grasped by the love of God – the notion of sin affirms a darker reality, namely our tendency to be drawn and dragged down.

We thus find ourselves excited and inspired by the vision of God, which draws us upwards; at the same time, we find ourselves pulled down by the frailty and fallenness of human nature. Earlier, we noted Alexander Pope's view that human beings were poised in the theological space between angels and beasts. The 'image of God' causes us to yearn to be angelic; sin inclines us towards our animal natures, driven by deep Darwinian instincts of survival, dominance and power. In a sense, Christian ethics is a principled refusal to conform to these social Darwinian principles, however pragmatically they are stated.[23]

Much the same point was made by Thomas H. Huxley in his 1893 Romanes Lecture at Oxford University, entitled 'Evolution and Ethics'.[24] Huxley noted that human animals have triumphed in the 'struggle for existence' through their cunning and 'ruthless and ferocious destructiveness'. Yet human beings, having subdued the remainder of the sentient world, now found that these 'deeply ingrained serviceable qualities have become defects'.[25] The violence and ruthlessness that secured their triumph over other animals were now seen as 'sins'; there was a recognition that the methods used in the 'struggle for existence' are 'not reconcilable with sound ethical principles'.[26] Ethics, for Huxley, is thus a principled resistance to precisely those animal qualities that secured human domination of the living world, and the Darwinian processes that underlie them. Yet – and here Huxley must be heard – this demands the subjugation of animal instincts that linger within us. Our hereditary history continues to shape our present – and it must be resisted, even though it cannot be eradicated. 'The practice of that which is ethically best – what we call goodness or virtue – involves a course of conduct which, in all respects, is opposed to that which leads to success in the cosmic struggle for existence.'[27] Evolution may explain the origins of ethics; it cannot itself function as the basis of ethics, in that we are now called to leave behind those former virtues of violence and aggressiveness which are today seen as vices. 'Evolution may teach us how the good and the evil tendencies of man may have come about; but, in itself, it is incompetent to furnish any better reason why what we call good is preferable to what we call evil than we had before.'[28]

Huxley's emphasis on the tension between hereditary forces that linger within us and our sense of justice and ethical principles finds echoes within the Christian tradition. Augustine of Hippo's basic criticism of Pelagius was his failure to allow for the continuing presence of inherited habits of thought and action, which are carried over into the life of faith. It is impossible to break free totally from our origins and contexts. Today, we would probably amplify his comments with reference to our evolutionary past.

The Christian 'big picture' depicts humanity as containing within itself a tension between two different modes of being. The New Testament uses the terms 'flesh (Greek: *sarx*)' and 'spirit (Greek: *pneuma*)' to refer not to parts of the human body, but rather to two different modes of human existence. Paul adopts what scholars term an 'aspective', rather than a 'partitive', understanding of terms such as 'soul', 'flesh' and 'body'.[29] To 'live according to the flesh' means to live at a purely human level, disregarding the spiritual side of life. Yet this theological framework can be extended to the moral and existential realms. It affirms that there is a tension within us as we feel drawn to good and to evil.

This framework emphasises the importance of divine grace, in that it recognises the limits placed on human beings for self-improvement and self-transcendence. We cannot heal ourselves; we need to be healed by someone else. We find ourselves drawn to sin, despite knowing that it is wrong; we find ourselves unable to achieve the good, despite knowing that this is what we should be doing. This dilemma was known and expressed by Paul: 'I do not do the good I want, but the evil I do not want is what I do' (Romans 7:19).

The Christian insistence that Jesus Christ is the saviour of humanity – not simply its teacher and moral example – reflects this view of the human situation. Salvation is not something that humans can achieve or earn; it is something achieved and given by God. Yet salvation is a process, in which a changed status in relation to God is gradually effected in terms of a transformed inner reality, often expressed using the language of 'renewal' or 'regeneration'. These do not designate an instantaneous change, in which someone suddenly breaks free from the power of sin to pull us down.

Augustine of Hippo used an analogy in making this point: we are like someone who is ill, but is receiving appropriate treatment, and is in the process of recovering. She is still ill; however, the healing process has begun, and she will eventually get better. She is thus both ill and better at the same time. She is ill in fact, in that she has not yet recovered, but she is healthy in hope, in that her healing is under way. Martin Luther famously expressed this notion when he declared that a Christian is 'at one and the same time a righteous person and a sinner (*simul iustus et peccator*).[30]

Christianity thus sets out a vision of human nature which recognises its enormous potential, while nevertheless affirming that it is wounded, damaged and broken. The New Testament uses the language of putting our old nature to death, and putting on a new nature – while at the same time recognising the lingering influence of our older selves in the life of faith. The 'Old Adam' lingers within us, shaping our motivations and actions (the parallels with Huxley's reflections on our hereditary history are worth exploring here). Human nature needs to be healed, and does not possess the capacity to heal itself. This framework helps us make sense of the complex picture we see of human culture and history, characterised by aspirations to greatness and goodness on the one hand, and oppression and violence on the other. From a Christian perspective, it is clear that we must recognise both the greater destiny of humanity and a corresponding capacity to fail to achieve such aspirations unaided and unhealed.

Individuals may indeed have the capacity for good; this seems matched, however, by their intrinsic capacity for evil. We are morally ambiguous, easily led astray. Recognising our own innate capacity for sin makes us less judgemental about others, in that we realise how easily we too might fall into such patterns of thoughts and behaviour. A recognition of this profound ambiguity and tension within us is essential if we are to avoid political and social utopianism, based on ideologically driven value judgements about human nature.

Science and Sin: Convergence without Identity

So where might science come into this? Nobody who knows anything about science or faith would be so crass as to suggest that their ideas on the human dilemma can be merged, or treated as if they are intellectually equivalent. We need to respect their quite distinct approaches and emphases. Yet the idea of 'enrichment of narratives' offers a way of respecting the individual identities of scientific and theological approaches, while at the same time exploring how their distinct perspectives might illuminate the complex reality that we call 'sin'. Many have turned to psychology for illumination of humanity's deeply puzzling tendency towards violence and self-destruction. Although the reliability of some of Sigmund Freud's judgements can easily be challenged, his own questioning of the myth of human goodness must be taken seriously. 'Men are not gentle creatures who want to be loved, and who at the most can defend themselves if they are attacked; they are, on the contrary, creatures among whose instinctual endowments is to be reckoned a powerful share of aggressiveness.'[31] The evidence strongly indicates that males show this tendency to violence significantly more than females, especially in the case of sexual violence.

Since the advent of Darwin's theory of natural selection, the imagery of conflict and a struggle for survival within the biological world has come to dominate discussion of human beings. Should we be surprised that humanity, having emerged from within this biological realm, is characterised by tendencies towards violence? Darwin's own language is highly suggestive here.

> The inevitable result is an ever-recurrent Struggle for Existence. It has truly been said that all nature is at war; the strongest ultimately prevail, the weakest fail . . . The severe and often-recurrent struggle for existence will determine that those variations, however slight, which are favourable shall be preserved or selected, and those which are unfavourable shall be destroyed.[32]

The development of related approaches within sociobiology also opens up some important possibilities for expanding our understanding of the nature of sin.[33]

We've already noted how both Thomas Huxley's lecture 'Evolution and Ethics' (1893) and Richard Dawkins' *Selfish Gene* (1976) opened up some fundamental questions about the biological origins of human tendencies towards violence and selfishness, and how this might be alleviated, if not entirely eliminated (151–152). The geneticist Steve Jones develops this point further,[34] reflecting on the dark side of human nature, and pointing to hereditary factors which might shape contemporary human attitudes. Traditionally, Christian theologians speak of 'original sin' in the sense of tendencies that lie within us from birth – rather than being acquired from our social context.[35] This resonates with genetic reflections on violence and self-centredness.

In her *Fall to Violence* (1994), Marjorie Suchocki considers sin in terms of instinctive human violence. 'A tendency toward aggression is built into human nature . . . we are by nature an aggressive species with a history of physical and psychic violence . . . The capacity for violence is built into our species through aggressive instincts related to survival. When that violence is unnecessary and avoidable, it is sin.'[36] This innate human tendency towards violence is expressed in multiple cultural forms, which extend far beyond the extreme case of international wars.[37] Yet it is important to notice that warfare does not represent a departure from basic human instincts, but actually represents their fulfilment.

Why are these scientific insights into human nature so important? In my view, their significance lies in their challenge to the naivity of the 'Age of Reason', which held that humanity was naturally good. The notion of original sin was rejected as insulting and demeaning to humanity, as it implied that human beings were born with some innate tendencies towards evil or irrationality. This, Enlightenment writers declared, was patently unfair. Yet we now know that we are born into this world with certain genetic dispositions, which may be further enhanced through our social context – just as they can be reduced. Jean-Jacques Rousseau famously declared that humanity is 'born free, and everywhere he is in chains' – implying that this imprisonment is the result of an unjust society, which must be overturned and reformed. The science of genetics suggests that we are not born 'free', but with inbuilt genetic tendencies which in themselves imprison us. Dawkins believes that we can overcome these by ourselves; I think we need help if we are to do so, and to transcend our natural limits and boundaries.

Sin: A Lens Through Which We See Ourselves

We are long past the historical moment when it was plausible to pretend that human beings are good. Earlier, I noted Anne Frank's heart-breaking confession that she held on to her belief that 'people are really good at heart', despite its obvious absurdity in the light of the Nazi occupation of the Netherlands (98). 'I simply can't build up my hopes on a foundation consisting of confusion, misery, and death.'[38] I make no criticism of her. We all need to 'tread softly' (W.B. Yeats) when walking over others' dreams. Yet I very much fear that hers is a dream that can only collapse and crumble in the face of a massive evidential dissonance.

We all need a lens through which we can see ourselves. Paradoxically, such a lens is something that we judge (is it right?), while at the same time being something which judges us (are we really like this?). We are often reluctant to be shown to be flawed, weak or foolish, and reject any such judgements primarily because they are threatening and embarrassing, irrespective of whether they might be true. Yet this difficulty is alleviated if we think of the lens as offering a diagnosis – an interpretation of the human condition.

For C.S. Lewis, Christianity provides a 'big picture' of reality which 'chimes in' with our experiences within us, and our observations of the world outside us. Lewis set out this vision in his book *Mere Christianity*, the opening chapters of which deal particularly with the human condition. Christianity offers an intellectual mirror, in which we see ourselves reflected and evaluated.[39] The resonance between what we 'see' in this mirror and what we experience of ourselves inclines us both to take this diagnostic capacity seriously, and to share in the hope of transformation that it brings. This is perhaps paradigmatically reflected in Paul's famous words in his letter to the Romans: 'Wretched man that I am! Who will rescue me from this body of death? Thanks be to God through Jesus Christ our Lord!' (Romans 7:24–5). Realism about our problems is thus suffused with hope about the future – a hope that is grounded in a faithful God, not in a misplaced optimism about ourselves or our world.

This analysis of human nature destroys any naïve optimism about human nature, demanding that we confront the darker forces that lie within us. Freud's identification of the darker shadowy forces at work within our unconsciousness may seem overstated at points; nevertheless, his point can hardly be ignored.[40] As Hannah Arendt pointed out, the unbearable horror of Auschwitz resides in the fact that its perpetrators were not Satanic or diabolical; they were 'human, all too human figures' trapped within all too human political and ideological systems.[41] What happened in the past could happen again.

So do 'human, all too human figures' stand at the centre of a nexus of evil, which they themselves have created, and which (other) human beings must oppose and dismantle? It is a most uncomfortable thought for those who insist on making humanity the defining frame of reference for morality – a view that many now term as 'humanism'. Yet, as we shall see in the next chapter, 'humanism' is a more elusive and complex notion than many realise. We must therefore look at the variety of humanisms, and consider how they deal with the moral ambiguity of humanity.

II

Humanisms: Secular and Religious

'Who then will not look with wonder upon man?'[1]
Giovanni Pico della Mirandola

As we reflect on the uncertain future of humanity, we need to ask what is distinctive and important about the human race. It is, at present, the most developed species in the natural world. Yet many are now reluctant to speak of humanity as representing the culmination of the evolutionary process. Why? Because, if Darwin is right, our presence at the top of the evolutionary tree is nothing but an accident, the consequence of a series of fortuitous and unintended twists in the unpredictable and uncontrollable narrative of evolution, doubtless helped along (as Thomas Huxley pointed out) by human cunning and violence. Charles Darwin may have found this an uncomfortable implication of his theory of natural selection, yet he did not hold back from sharing it with his readers: humanity's elevated position within the order of nature was an undeserved fluke.[2]

We therefore need to think about *humanism* – an attempt to understand what is distinct and good about human beings, and how such virtues can be safeguarded. In the lazy and uncritical world of twenty-first-century journalism, the meaning of the word 'humanism' is self-evident and unproblematic. It designates ways of thinking and living which exclude God, focusing instead on the achievements and aspirations of human beings. The idea of a 'Christian humanism' is thus dismissed as an oxymoron, a contradiction in terms, one of the impossible things that Lewis Carroll's Alice valiantly tried to believe before breakfast.

Yet this influential way of thinking about humanism cannot be sustained without doing violence to contemporary experience or to history. It is not exactly difficult to point to a very large group of human beings who frame their aspirations and ground their achievements in religious terms! The real difficulty, however, is historical, in that this specific concept of *secular* humanism – to give it its proper name – was invented in the twentieth century. It bears little relation to the nobler and more generous visions of humanism that emerged in the European Renaissance, associated with genial and wise writers such as Erasmus of Rotterdam. We need to reclaim such an older and wiser vision of humanism, which is much better adapted for the challenges that humanity faces in the twenty-first century, and is more rigorously grounded in history and science than the impoverished 'secular humanism' that has temporarily gained the ascendancy. Let's begin by doing some history.

The Renaissance: The Historical Origins of Humanism

Humanism was the intellectual glory of the Renaissance, that remarkable period in European history from about 1300 to 1600 which witnessed the transformation of culture through an immersion in the cultural vitality of the ancient world. It was a programme without parallel at the time, attracting individuals on account of the commanding power of its aesthetic vision, and its sense of new possibilities that would enrich the lives of individuals and nations. Humanism arose in Italy in the 1300s, and slowly made its way northwards and westwards, becoming a significant presence in England in the early sixteenth century.

My own interest in Renaissance humanism developed during the 1980s, when I specialised in a group of humanist writers based in Switzerland, including Joachim von Watt (1484–1551), Heinrich Glarean (1488–1563), and Johannes Xylotectus (1507–68). These writers did not fit into a rigid intellectual template, and exemplified the diversity that is now generally regarded as characteristic of Renaissance humanism. As I worked in university archives in the cities of Zurich, Vienna and St Gallen,

I began to absorb something of the excitement which saturated the cultural and ecclesial atmosphere of the age.[3] Renewal and regeneration were in the air! It was possible to step over the 'Middle Ages' (a phrase, by the way, invented by supporters of the Renaissance) which separated the present from the glories of antiquity, and allow their fruitful reconnection.

Modern readers tend to assume that 'humanists' were individuals who subscribed to a common body of beliefs, attitudes and values known as 'humanism', in much the same way as Marxists are individuals who subscribe to the ideas of Marxism. Yet there is little historical evidence for this assumption. Renaissance humanism was not an ideological programme of any kind, still less an anti-religious movement.[4] It was fundamentally about the 'pursuit of eloquence' – the renewal of a moribund culture through the 'revival of good letters'.[5]

Referring to someone as a 'humanist' in the fifteenth or early sixteenth century tells us little about their philosophical, political or religious views. In fact, the term 'humanism' is best seen as a nineteenth-century invention. During the time of the Renaissance itself, the word 'humanist' (Italian *umanista*) was widely used to refer to a university teacher of *studia humanitatis* – a Latin phrase designating 'humane studies', or the 'liberal arts', such as poetry, grammar and rhetoric. The English word 'humanist' first appears in 1589, with the sense of 'a literary scholar, especially someone versed in Latin studies'.

The Italian Renaissance is so multi-faceted that just about every generalisation concerning its characteristic ideas tends to be a distortion. It is for this reason that the view of humanism developed by Paul Oskar Kristeller (1905–99) is of such importance, in that it was able to account for at least a substantial amount of the remarkable diversity of outlooks evident within the Renaissance.[6]

Kristeller offered a nuanced and sensitive account of humanism, which envisaged the movement as essentially cultural and educational, primarily concerned with written and spoken eloquence, and only secondarily concerned with matters of philosophy and politics. Although Renaissance humanism was characterised by no distinctive philosophical or ideological stance, the fact remains that, virtually without exception, Renaissance humanists were Christians operating within the context of the life and thought of the Church, and concerned for its reform and renewal. The Enlightenment tendency to portray the humanists as precursors of its critique of religion lacks plausibility, not least because leading humanists of the age – such as Giovanni Pico della Mirandola, Lorenzo Valla and Desiderius Erasmus– tended to see humanism as continuous with the medieval Catholic spiritual tradition, rather than as a precursor of rationalism.

In debates about the 'New Atheism' between 2006 and 2011, I was astonished to find that some followers of this movement seriously believed that Erasmus and other leading Renaissance humanists were atheists. When I asked them for documentary evidence of this, they declared that it was below their dignity to respond to such ridiculous questions. If these writers were humanists, then by definition they were atheists. When I asked them to explain Erasmus's strongly religious views in works such as his New Testament commentaries, or his highly influential *Enchiridion Militis Christiani* ('Handbook of the Christian Soldier', 1503), it became clear that they had not read any of them, and were dependent on popular atheist writers for their ridiculous views. I began to realise it was necessary to challenge this nonsense. Renaissance humanism has, to put it bluntly, been hijacked in the service of modern ideological agendas. Truth, as they say, is the first casualty of warfare.

So how did this myth of an intrinsically atheist humanism take root? In fact, this turns out to be a recent development, in which a specific and unrepresentative form of humanism – best designated as 'secular humanism' – asserted proprietorial rights over the more dignified and noble views of the Renaissance. Paul Kurtz, one of America's most prominent sceptics and atheists,[7] played a leading role in reshaping and redirecting American humanism in a specifically secular fashion during the late 1970s and early 1980s, largely by suppressing both its historic religious origins and continuing religious associations and commitments. While the original American 'Humanist Manifesto' (1933) made specific approving reference to religious humanism,[8] Kurtz vigorously advocated more secular forms of humanism, and founded the 'Council for Secular Humanism' to lobby for a more aggressively anti-religious attitude within the American Humanist Association. He was one of the two primary authors of 'Humanist Manifesto II' (1973), setting out a vision for humanism framed exclusively in secularising terms.

Lazy and uncritical media reporting ensured that Kurtz's vision of *secular* humanism quickly became elided with humanism *as a whole*, and was read back into older visions and implementations of the movement. This process gave rise to the utterly incredible and unevidenced belief that Renaissance humanism was an essentially atheistic movement. This makes a nonsense of history – at least, to those who know that history and realise the intellectual violence that is being done to historical scholarship by ideologically driven agendas. Although serious disagreements within the Council for Secular Humanism made headlines in 2010,[9] media reporting of humanism continues to be trapped within a secularist framework. It's time to move on from this outdated and indefensible perception.

Readers of this book who are unaware of the rich intellectual and cultural heritage of the Renaissance need to know that the term 'humanism', as used today, bears little relation to its historical predecessors. Ludwig Wittgenstein once remarked that certain terms needed 'to be withdrawn from language and sent for cleaning' before they can be 'put back into general circulation'.[10] It is hard not to appreciate the wisdom of his remark for our topic. So can the term 'humanism' be cleaned up and put back into general circulation, shorn of distorted interpretations and sectarian agendas? It's time to recover the word from the ideologues and put it back into circulation for everyone to own and use. There is, and never has been, any problem in speaking of 'Christian humanism' – save on the part of those concerned to impose their dogmatic and sectarian understanding of 'humanism' on everyone else.

Whatever else humanism may be – and there are multiple understandings of its significance, some secular and some religious – all forms of humanism affirm the importance of humanity, and its distinctiveness within the natural order. As the great Renaissance writer Pico della Mirandola remarked, 'Who then will not look with wonder upon man?'[11] The rise of the 'Age of Ideology' in the twentieth century saw the rise of dehumanising ideologies, such as Nazism and Stalinism. The forceful reassertion of a theologically grounded view of human rights and dignity in the Soviet bloc – then teetering on the brink of collapse – was one of the most significant achievements of Pope John Paul II, who spent much of his life working in Poland during the Soviet era.[12] This vision of Christian humanism, grounded in a robust theological understanding of the human person, may only be one version of humanism – but it serves both to demonstrate the plurality of humanisms and to undermine the claims of any one variant of humanism to speak for the entire movement.

Yet there is another issue that emerges from this brief discussion of John Paul II's religious humanism. Does the existence of religion itself undermine or subvert the entire intellectual and moral plausibility of secular humanism? In what follows, we shall pick up on a concern expressed in the writings of the great Oxford philosopher Bernard Williams (1929–2003) – himself a secular humanist – that the movement has never really faced up to the intellectual and moral consequences of its belief that religion is a human invention. It's an important point, and we shall consider it in some detail.

The Problem of Religion for Secular Humanism

As a secular humanist, Williams had a marked distaste for religion, but he was acutely aware of the intellectual implications of any suggestion that religion was evil or degenerate. If there is no God, or any transcendent being, then religion is a mirror, reflecting the heart and soul of its creators – human beings. And that, he pointed out, exposed some glaring internal contradictions within 'humanist' thinking.

> For granted that [religion's] transcendental claim is false, human beings must have dreamed it, and we need an understanding of why this was the content of their dream. (Humanism – in the contemporary sense of a secularist and anti-religious movement – seems seldom to have faced fully a very immediate consequence of its own views; that this terrible thing, religion, is a *human* creation.)[13]

So if religion is indeed intrinsically degenerate and perverting (a belief that Williams regarded as normative), this ought to give humanists grave cause for concern about the goodness and integrity of human nature itself.

This is an important point, and needs to be explored thoroughly. The New Atheists hold that God is to be blamed for the evil of the world. God is nasty, vindictive and oppressive. This fundamental criticism is summed up by Richard Dawkins in an oft-quoted, if slightly hysterical, passage from his *God Delusion*: 'The God of the Old Testament is arguably the most unpleasant character in all fiction: jealous and proud of it; a petty, unjust, unforgiving control-freak; a vindictive, bloodthirsty ethnic cleanser; a misogynistic, homophobic, racist, infanticidal, genocidal, filicidal, pestilential, megalomaniacal, sadomasochistic, capriciously malevolent bully.'[14] Yet this New Atheism scapegoating of God for the rational and moral failings of human beings merely exposes what many, including Williams, see as a fatal contradiction within its own worldview. Everything that is wrong with the world, the New Atheism assures us, can be blamed on God. But if God is an invention, a 'fictional' character with no existence outside deluded human minds, that idea of God can only emerge from human minds. Christopher Hitchens argues that human beings create God in their own likeness, attributing their own moral and rational qualities to these supposed supernatural beings. 'God did not create man in his own image. Evidently, it was the other way about.'[15] Both God and religions must be recognised to be 'man-made'.[16]

But there is clearly a problem here. The logic of both Dawkins' and Hitchens' analysis of God and religion as human fabrications leads to the conclusion not that religion corrupts an innocent humanity, but that corrupt human beings create a religion that is just as evil and degenerate as they are. Let me playfully rewrite Dawkins' piece in the light of this point.

> The God of the Old Testament is arguably the most unpleasant character in all fiction, created by equally unpleasant human beings who were jealous and proud of it; who were petty, unjust, unforgiving control-freaks; who were vindictive, bloodthirsty ethnic cleansers; who were misogynistic, homophobic, racist, infanticidal, genocidal, filicidal, pestilential, megalomaniacal, sadomasochistic, capriciously malevolent bullies; and who created their gods in their own image.

More aggressive forms of secular humanism – such as the New Atheism – are thus caught in a dilemma which is both framed and created by two of its core beliefs: God is evil and nasty; and God is a delusion created by human beings. The more the New Atheism excoriates religion as irrational and immoral, the more it inadvertently highlights the irrationality and immorality of its human creators. This is obviously a problem for secular humanism. As Williams points out, the root of the problem is not religion, but humanity. If Dawkins is right, and God and religion are just nasty human delusions, what does this tell us about the moral and intellectual status of the human beings who invented these nasty ideas?

The easiest way of getting out of this mess is to develop a New Atheist variant of Manichean dualism, which declares that some human beings are moral and rational (and hence secular), and others are evil and intellectually degenerate (and hence religious). This strategy lies behind the use of the word 'Bright' as a positive designation of someone with secular or atheist views. It was enthusiastically commended by Daniel Dennett and Richard Dawkins back in 2003.[17] Brights weren't responsible for the outrageous invention of God; both God and religion were dreamed up by less intelligent, less enlightened human beings. Yet the notion of the 'Bright' was widely ridiculed – even within secularist circles – as elitist and self-congratulatory. Christopher Hitchens openly mocked Dawkins and Dennett for their 'cringe-making proposal that atheists should conceitedly nominate themselves to be called "Brights" '.[18]

For most of us, religion is a striking example of something with great potential for good but that can go badly wrong. William Temple (1881–1944), sometime Archbishop of Canterbury, delivered a series of lectures at the University of Glasgow on the theme of religion in 1932, in which he declared that there were good reasons for suggesting that religion had done more harm than good to humanity. The problem lay in the corruption of religion. 'Religion itself, when developed to real maturity, knows quite well that the first object of its condemnation is bad Religion, which is a totally different thing from irreligion, and can be a very much worse thing.'[19]

Nobody denies that religion can go badly wrong. Yet religion here turns out to be a particularly luminous example of a general problem with human nature. Just about everything that we blame for enslaving or perverting us turns out to be a human creation. Terry Eagleton thus takes no small pleasure in lampooning the self-indulgent and self-congratulatory optimism of much Western liberalism by pointing out that the social evils against which it pits itself are themselves the outcome of human decisions and actions. We blame society for our ills, but just who created that society in the first place?[20] Human beings thus act as both their own oppressors and liberators.

As Aristotle pointed out, human beings are social animals, who create societies – and those societies are shaped by human values. Reinhold Niebuhr argued that, while individuals often showed a willingness and ability to transcend self-interest in their personal dealings, this tendency was at best imperfectly realised at a corporate level. Relations between nations, ethnic groups or social classes often show little if any capacity for self-transcendence. Yet paradoxically, a dehumanising society is ultimately a human creation, a depressing witness to both the fragility of human moral concerns and the lingering power of human self-interest.[21]

So what can be done to change this situation? To make both individual human beings and human societies more just and humane? For many, the answer lies in education.

Humanism and Education

Renaissance humanism regarded education as essential to the future of humanity. It was a channel for the transmission of inter-generational wisdom, capable of creating a moral and cultural vision for the future of humanity. My own study of Swiss humanism in the early 1500s brought home to me how much education mattered to the Renaissance. Many humanists of the age saw the study of the humanities – such as history and rhetoric – as keys that unlocked the potential of human beings to achieve their true potential.

It is a bold and brilliant vision, characteristic of Renaissance humanism at its best. Yet some will wonder whether it quite matches the remarkable capacity of human beings to do utterly stupid and destructive things. Can we be educated to be good? Or do we need some kind of moral transformation in order to help us to love good in the first place, and pursue it in the second? Some within the humanist movement of the Renaissance took the view that only God is capable of bringing about the radical transformation of human nature that is needed if we are indeed to love good, rather than our own self-interests. Others hoped for a rationalisation of self-interest, limiting its toxic scope.

It's a long way from the upbeat optimism of the Renaissance to the year 1942, when Stanford University in California opened its new School of the Humanities. The celebration of that event was somewhat muted. How could the cultural achievements of humanity be applauded and promoted as a force for good in the light of the devastation of the Great War (1914–18), and the rise of totalitarianism in many parts of Europe previously regarded as bastions of civilisation? As if that were not enough, the United States itself was now caught up in a new global conflict, having declared war on Japan and its allies a few months earlier, following the bombing of Pearl Harbor in December 1941. John W. Dodds (1902–89), the first Dean of the School of Humanities, conceded it was not a particularly propitious moment to celebrate human cultural achievements. Why establish 'an outpost of the humanities' and talk about the 'place of culture in our civilization' when that civilisation itself was teetering on the brink of disaster?[22]

Dodds acknowledged the force of this point. The disasters into which the world had lurched in recent decades seemed to speak more of inhumanity and insanity than of the great human cultural virtues. Tempering his idealism with a heady dose of realism, Dodds offered a sober account of the enigmas and inconsistencies of human nature, above all the need to confront our capacity for destruction.

[Human technological] development has been accompanied by a progressive dehumanization of society, until at last the Frankenstein monster seems in a good way to wipe out its master. Man's mechanical contrivances are the marvel of the world, but morally and ethically and socially we are still (relatively speaking) trying to crawl out of the primeval slime. Man has gained mastery of his environment but he seems to be less and less master of himself. Today we see him turning the weapons of his brain against himself—groping, amid the noise of a tottering civilization, for some faith in man to which he can cling.

Yet for Dodds, the humanities articulated a vision which might both sustain a nation during a time of war, and guide its post-war construction. 'We need to be reminded of the dignity of man, to remember that we are potentially noble, to learn the infinite worth of the individual.' Once the war was over, it might be possible to recover such lost ideals and dreams.

For me, some sections of Dodds' 1942 speech stand out as being among the finest defences I have read of the place of the humanities in safeguarding certain core values of Western culture, and of the role of education in creating better and wiser people. Yet there are points at which his speech seems so unrealistic about the moral qualities of humanity that it verges on the utopian. There has always been a stream of thought within Western culture which holds that education transforms people, converting them from a selfish ignorance to a generous toleration. Yet this often reflects the importance of peer pressure in maintaining positive and generous social values, and the social role of universities in influencing students' political and social views through judicious selection processes, which filter out those with undesirable attitudes and outlooks. It is unclear whether education itself – as opposed to other social processes – modifies the human tendency towards self-interest and self-promotion.

The outbreak of the Second World War precipitated a fresh outburst of reflection on the enigma of human nature. Although the full horror of the Nazi extermination camps did not become apparent until 1945, as the war in Europe moved into its final phases, many suspected that both sides of the conflict had committed atrocities that could only be described as 'inhuman'. So how could human beings behave in such an inhuman manner? Surely a good education would prevent such things from happening?

Maybe it should have. But it didn't. One example will suffice to indicate the scale of the problem. In January of 1942, thirteen Nazi technocrats gathered for a conference at 56–58 am Grossen Wannsee, a villa in a leafy suburb of Berlin.[23] Most of those present were highly educated, with doctorates or medical qualifications from leading German universities. Their task was to agree protocols and procedures for the elimination of Jews from Germany and the occupied territories. One copy of the protocols agreed by the meeting has survived, detailing – although in somewhat euphemistic language – the arrangements for the deportation, economic exploitation and final extermination of the Jews.[24] While I like to think that learning makes us into better people, I have to admit that I find it very hard to see how their education 'humanised' these people.

One of the more thoughtful reflections on such themes came from the pen of C.S. Lewis, whose fame in Britain stemmed initially from a series of wartime talks, broadcast by the British Broadcasting Corporation.[25] While realising that humanity's capacity for self-deception was so great that such sombre lessons would simply be suppressed and forgotten once the war was over, Lewis nevertheless took the view that optimistic ideas of human nature had been dealt a deathblow by the horrors of the war. Lewis's *Abolition of Man* (1943) is one of his most difficult books, lacking the clarity of many of his works of this period. Yet alongside Aldous Huxley's *Brave New World* (1932), it remains one of the most eerily prescient books of the twentieth century, opening up awkward questions about human nature that just will not go away – including our failure to learn from our past.

The Threat of Dehumanisation

One of the core tasks of any viable form of humanism – whether Christian or secular – is to resist the perennial threat of dehumanisation. This almost invariably takes the form of one social 'in-group' declaring that a social 'out-group' is less than human

– for example, in referring to its members as 'animals' or 'vermin'. The most familiar example of this is the dehumanising racial ideology of National Socialism, which treated Jews, gypsies, Poles and Serbs as *Untermenschen* (German: 'sub-humans'). Yet recent history is littered with genocides which reflect the influence of dehumanising ideologies – such as the Armenian genocide of 1915–17, or the Rwandan genocide of 1994.[26] Part of the solution must be to challenge and neutralise such value-laden ideologies, which cause one set of human beings to regard others as representing a lower form of life.

Yet the threat of dehumanisation is broader than this. It lurks within the tendency to commodify humanity, treating persons as if they were objects of potential commercial value or economic exploitation. This corrosive process of depersonalisation leads to certain individual humans being treated as an 'It' rather than as a 'You'. This trend can be seen in the Nazi genocides. Individuals were first robbed of their valuables, then stripped of one of the most fundamental marks of human dignity and identity: their names. Now anonymous, these extermination camp inmates were known only by the numbers tattooed on their arms. Finally, they were treated as economic assets, being stripped of their hair and other valuable body parts, before being liquidated in an industrial-scale killing process. There is simply no parallel to this within the animal realm. Human beings alone seem morally and physically capable of genocide.

Ideologies both generate and sustain this dehumanising outlook. Sometimes these ideologies are religious, with some groups regarding themselves as 'God's chosen people'. Yet we must not overlook ethnic ideologies which speak of the 'Master Race', or those which dehumanise older people on account of their failing mental or physical capacities. The Nazi 'Action T4' programme of 1939–41 – which led to the involuntary euthanasia of some 70,000 psychiatric patients – was based on the assumption that such people were simply devoid of any value or utility. It provoked a powerful reaction from Pope Pius XII in 1943:

We believe that it is necessary to reiterate this solemn statement today, when to our profound grief we see at times the deformed, the insane, and those suffering from hereditary disease deprived of their lives, as though they were a useless burden to society; and this procedure is applauded by some as a mark of human progress, and as something that is entirely in agreement with the common good. Yet who that is possessed of sound judgment does not recognize that this not only violates the natural and the divine law written in the heart of every person, but that it violates the noblest instincts of humanity?[27]

While there are many issues that need to be addressed, one of the most pressing is the need to develop a way of seeing humanity which is capable of resisting the denigration or devaluation of any groups of people, or isolated individuals. The preaching of Jesus of Nazareth was not about the creation of a new 'in-group', but about the subversion of 'in-groups' in general. So what theological foundation might be offered for this principled repression of such a fundamental human social instinct leading to the creation of 'in-groups', which is as biologically natural as it is socially damaging? One answer lies in the notion of humanity being created in – and still bearing – the 'image of God'.[28]

The Christian belief in humanity bearing the 'image of God' affirms that humanity has an identity and value which transcends our economic utility and social location. One of the earlier discussions of this within the Christian tradition is due to Lactantius (c.240–c.320), who argued that the belief that human beings are created in the 'image of God' demands equality among peoples, and gives rise to affection and respect across social and political barriers.

Now it was from the one human being that God created us all, so that we are all of the same blood, with the result that the greatest crime is to hate humanity or do them harm. That is why we are forbidden to develop or to encourage hatred. So if we are the work of the same God, what else are we but brothers and sisters?[29]

For Lactantius, this theology implied that slavery was unacceptable, since all human beings were created equal, bearing God's image.[30] Few were bold enough to follow him, yet the theological logic of his position is irresistible.

Others would not agree. If humanity is prone to weakness and the degeneration of old age, why not take charge of the evolutionary process, and bring about a radical change in human nature, so that these problems might be consigned to the past? In the next chapter, we shall consider some approaches to this question, including the movement known as 'transhumanism', which argues for the transformation of humanity through technological progress.

12

The Myth of Progress: Reshaping Humanity

'Amid arts forgotten, commerce annihilated, fragmentary
literatures and populations destroyed, the European
talks of progress because by an ingenious application
of some scientific acquirements he has established a
society which has mistaken comfort for civilization.'[1]
Benjamin Disraeli

The dominant narrative of Western culture is that of *progress*.
Things are not simply changing; they are getting better. We thus
move from an undeniable empirical observation to a contestable
value judgement, which has assumed the status of an unques-
tioned and unquestionable dogma, serving functions that are in
some ways similar to those of religious beliefs.[2] This all too easily
collapses into the Panglossian view that all change is good, no
matter what is being changed or who is changing things. Anyone
who suggests otherwise is dismissed as wallowing in nostalgia,
longing wistfully for a misremembered and idealised past.

Any attempt to explore these issues needs to engage with the
question of what the word 'progress' means. Most understandings of progress incorporate three core themes.[3]

1. Improvements in objectively measurable qualities, such as life
   expectancy, perinatal mortality or gross domestic product.
2. An improvement in the human situation, such as the growth
   of virtues such as tolerance, freedom or equality. (This, of
   course, depends on a consensus about which of these non-
   empirical qualities actually count as 'virtues'.)
3. An improved understanding of our world, particularly
   through scientific progress, as a result of which our ways of

depicting and speaking about the world become more reliable.

None of these can be considered adequate on its own as a measure of progress, and they are best seen as threads that are to be woven into a greater narrative.

Yet there are difficulties here. Perhaps the most challenging is that 'progress' is an ideologically freighted notion, not a simple description of events. We *observe* that things change; we *judge* that these changes are progressive or regressive. A development which seems regressive within a moral framework of one ideology appears as progressive within another. This can be seen from English reactions to the French Revolution of 1789. Some people, like the young Wordsworth, saw this as socially transformative and liberating.

> Bliss was it in that dawn to be alive,
> But to be young was very heaven![4]

Others at the time, however, saw it as leading to social chaos and instability, threatening the future. Progress is the way we refer to change that we deem to be good. But just who are 'we'? All too often, 'we' refers to the judgements of some privileged and powerful elite, who presume to speak for all right-thinking people, and tell the rest of us what we ought to think. In the end, whether something or someone is 'progressive' is determined by what the philosopher John Dewey termed 'ends-in-view' – the goals that people believe to be desirable.[5]

The ideal of 'progress' is clearly dependent on certain assumed moral values. It is not an empirical notion (such as 'change'), but is rather an evaluation of the merits and significance of these changes. These values are determined by interest groups. The importance of this point is easily grasped when considering the colonial expansion of European nations in the late eighteenth and nineteenth centuries.[6] European colonial powers all too easily thought of 'progress' as civilising savages through benevolently imposing their own self-evidently correct set of moral and social values on conquered cultures, implicitly regarding this as cultural 'progress'.

Certain groups of thinkers define themselves as 'progressive'. It is, however, a notoriously ambivalent term, grounded in some conspiratorially unspoken and unacknowledged understanding of how history *should* develop. This is particularly clear within both Marxism and Marxism-Leninism, conceptually obese systems which try to force every historical event into some theoretical hole. Both these bloated worldviews held the essentially determinist belief that the inevitable outcome of the historical process was the triumph of socialism. The difficulty with Marxism was not so much that it was an 'elaborately sophisticated structure erected on the foundation of a primitive misconception',[7] but that it rested on the precarious thesis of inevitable historical 'progress' towards a predetermined and knowable goal, which history itself seemed unable to confirm.

Given this ideological precommitment to the historical inevitability of socialism, it was natural for Marxists to speak of the 'progressive forces of socialism and the degenerate forces of capitalism'.[8] Progressives were those who went with the flow of history, reactionaries those who sought to prevent history reaching its intended goal, or even tried to reverse the historical process through counter-revolution. The difficulty, however, was that history itself did not seem to know about the rules that governed its supposed unfolding.[9] And furthermore, just why should the inevitable be deemed to be *right*?

The theologian Reinhold Niebuhr rightly pointed out the moral and intellectual vulnerability of liberal theories of progress, which so often seemed to elicit admiration from those who hoped that they were right, despite their blatant failure to face up to the human potential to choose and create evil.

The idea of progress is the underlying presupposition of what may be broadly defined as 'liberal' culture. If that assumption is challenged, the whole structure of meaning in the liberal world is imperilled. For this reason the liberal world is intolerant in regard to this article of its creed. It does not argue about its validity, precisely because it has lost every degree of scepticism in regard to it.[10]

Yet, as Niebuhr pointed out, this liberal 'creed' was 'highly dubious', not least because 'all historical processes are ambiguous'.

Now Marx may have chosen the wrong goal. Yet with crystalline acuity, he realised that progress is a *teleological* notion. It is about identifying the right goal, and doing what is necessary to move ourselves and others towards it. If progress, though, is indeed about attaining a goal, it must include a critical capacity for reflection and a willingness to change direction if the evidence suggests that our present trajectory is not bringing us closer to that goal – or that we have chosen the wrong goal in the first place. It's an obvious point, made with characteristic clarity by C.S. Lewis:

> Progress means getting nearer to the place you want to be. And if you have taken a wrong turning, then to go forward does not get you any nearer. If you are on the wrong road, progress means doing an about-turn and walking back to the right road . . . I think if you look at the present state of the world, it is pretty plain that humanity has been making some big mistake. We are on the wrong road. And if that is so, we must go back. Going back is the quickest way on.[11]

The Need for Realism, Not Optimism

It is becoming much more difficult to sustain an uncritical metanarrative of progress in the West, as the manifest failings of Western liberalism have become increasingly clear.[12] This inadequate metanarrative is one of the chief targets of Terry Eagleton's withering critique of the 'New Atheism', which he regards as uncritically reflecting this 'myth of progress'.[13] Eagleton regards the 'dream of untrammelled human progress' as a 'bright-eyed superstition' which lacks any rigorous evidential base. 'If ever there was a pious myth and a piece of credulous superstition, it is the liberal-rationalist belief that, apart from a few wrong turns, we are all steadily en route to a finer world.'[14] Eagleton dismisses this myth as a demonstrably false pastiche, a luminous example of an infantile 'blind faith'. What rational thinker could conceivably sign up to such a secular myth, which is obliged to treat such human-created catastrophes as Hiroshima, Auschwitz and apartheid as 'a few local hiccups' which in no way disrupted the steady upward progress of history?[15]

Underlying this 'myth of progress', Eagleton argued, was a naïve optimism which sorely needs to be challenged and reconstructed. For a start, we need to acknowledge the presence of a powerful and thoroughly authoritarian current in Enlightenment thought, evident in more radical elements of the French Revolution, which supported and practised methodical violence as a means of improving society. A classic example of this can be found in Eugene Lyons's account of a period he spent in the Soviet Union during the 1930s, when repressive measures were defended on the basis of the degree of economic progress they helped to facilitate. 'Every new statistical success gave another justification for the coercive policies by which it was achieved.'[16] The first and greatest casualty of this obsession with optimism is not so much truth, but any attempt to be realistic. 'Reality is a pessimist to whose treasonable talk one must shut one's ears. Since the truth is often enough unpleasant, it must be trumped by the unflinching will. It is a vein of optimism not easy to distinguish from mental illness.'[17]

For Eagleton, one of the most pressing difficulties with this optimistic view of human progress is the troubling persistence of evil.[18] The cosmic optimism of the eighteenth century proclaimed the rationality and harmony of the universe, and the elevated position of human beings as rational interpreters and inhabitants of this ordered paradise. Although indifferent towards theology, the problem that arose from this naïve cosmic optimism was thoroughly theological: how can evil arise within paradise? 'Such cosmic optimism tends to be self-defeating, since it throws into relief what it finds hardest to accommodate.' The difficulty, in my view, can be traced back to Descartes and the origins of the Enlightenment. Descartes' unwise emphasis on divine perfection merely accentuated the dissonance between theory and observation,[19] creating an intellectual crisis where previously there had been merely some slight mental discomfort, which was perfectly capable of being accommodated within its leading metanarratives.

Thinkers of the nineteenth century, Eagleton suggests, adopted a somewhat different approach to the problem of evil. Though the *bien pensants* of that age were prepared to admit that evil was clearly an experienced reality within the world, they nevertheless countered that evil was in the process of being eliminated through human progress. This served the purpose of holding realism and optimism together in a rather delicate balance, allowing the irritating presence of evil to be conceded, while accentuating the upbeat and optimistic belief that evil was well on the way to being eradicated by an enhanced human wisdom and knowledge, and technological capacity to alter our environment. Humanity would ultimately triumph over evil! Curiously, the idea that humanity might be the agent – and not merely the victim – of evil remained largely unexamined by such enthusiastic social progressives.

To defeat evil, we need to triumph over ourselves. Human beings create and cause evil. It makes no sense to speak of an earthquake as a 'natural evil'. It's just a natural process, which happens to have unfortunate implications for human beings who are unfortunate enough to be in the wrong place at the wrong time. But it makes perfect sense to speak of human evil – as seen in the great genocides of the twentieth century, which have no parallels in the animal world. If we are to avoid evil, we need to change. So how might that happen? In what follows, we'll look at some proposed solutions and their potential downsides.

Technology and Progress

The Enlightenment's narrative of progress affirms the improvement of human society and social conditions through advances in scientific understanding and technological application. Reason and science lead to the progressive improvement of the human condition through the erosion of religious superstition, and the emancipation of humanity from all taboos and arbitrary limits. The coupling of technological development and social progress was anticipated in the writings of Francis Bacon (1561–1626),[20] and remains a dominant theme in modern thinking on progress. Scientific advance is inextricably linked to social progress.

Now let's be clear. In many ways, human beings have improved the conditions under which they live. Medical advances have unquestionably led to an extension of life expectancy. Diseases that once wiped out populations can now be controlled. Technological advances have led to greater crop yields, thus increasing the world's food resources. Yet these are always part of a complex picture. Just as a conveniently selective attention to only part of this picture creates the semblance of uniform progress, focusing on other parts creates the equally unreliable appearance of collapse and degeneration. Advances in one area mask retreats in others. Technology has changed things for the better – and for the worse.

The history of the twentieth century is perhaps the greatest obstacle that the metanarrative of secular progress has to overcome, not least because of its sacred mythology of the unique capacity of religion to generate violence. The First World War, the Great Depression and the Second World War all raised awkward questions about the plausibility of this narrative. We were told that if we got rid of religion – or at least neutralised it, pulling out its teeth – then the likelihood of war would be drastically reduced, since religion was a key element in causing global conflict. Yet as far as scholars can see, there were no significant religious motivations for either the First World War (death toll around 16 million) or the Second World War (death toll around 60 million). We need to face up to the fact that, as a species, human beings are animals that use violence to achieve their ends, and technology to extend the reach of that violence. Religion doesn't cause this tendency; it merely reflects it.

That was certainly the judgement of J.R.R. Tolkien, who served as a British infantry officer during the First World War. He experienced there the terrifying reality of modern technological warfare, in which machines hurled an impersonal death from a distance towards their targets.[21] Tolkien came to see the 'machine' as a symbol of the lust for power – above all, for power over others. Where some might see machines as devices that liberate people from servility, Tolkien saw them not as a means of liberation but rather as tools of coercion, domination and enslavement.[22]

Some will rightly suggest that this is an excessively negative view, reflecting Tolkien's traumatic experiences of a technologically enhanced warfare. Yet we cannot entirely dismiss his concerns. In his *The Lord of the Rings*, Tolkien contrasts the hobbits, who used simple technology to help them stay close to the natural world and enjoy a simple life, with human beings, who used technology as a means of domination and exploitation.

Artificial Breeding: Eugenics and the Human Gene Pool

Charles Darwin's *Origin of Species* (1859) argued that the standard stockbreeding practice of 'artificial selection' offers an insight into a corresponding process which takes place within the world of nature.[23] Darwin extended his analysis in *The Descent of Man* (1871), arguing that human beings are also an outcome of this process of 'natural selection'. It was not long before some leading figures in Victorian culture, anxious to safeguard both the future of the human race in general, and the cultural supremacy of Great Britain in particular, suggested that the same principles used in stockbreeding should be applied to human beings. Why not breed humans selectively, to ensure that those with desirable physical or temperamental qualities dominated?

Darwin himself laid the foundations for this development. In his *Descent of Man*, he noted that primitive societies rapidly eliminated those who were 'weak in body or mind', so that those who survived were more healthy. For example, the ancient historian Plutarch told of how the city of Sparta identified weak infants, and eliminated them as a potential liability for the city.[24] Civilised societies, Darwin remarked, hindered this 'process of elimination' through medical and social care, thus enabling 'the weak members of civilised societies' to reproduce – with negative implications for the health of future generations. Remedial action was clearly required.

> No one who has attended to the breeding of domestic animals will doubt that this must be highly injurious to the race of man. It is surprising how soon a want of care, or care wrongly directed, leads to the degeneration of a domestic race; but excepting in the case of man himself, hardly any one is so ignorant as to allow his worst animals to breed.[25]

Darwin here arguably offered nothing more than implicit support for the notion of selective breeding for humans, paralleling the best practices of stockbreeders. Others, however – such as Sir Francis Galton (1822–1911), a cousin of Darwin – saw how Darwin's ideas opened the way to the preservation of the human race through selective human breeding.[26] For the future good of the human race, certain types of human beings ought not to be allowed to reproduce.

Although we tend to think of such repressive policies as characteristic only of a degenerate Nazi Germany, they were in fact widely accepted within progressive circles in America and England during the 1920s. In fact, it is difficult to identify anything about eugenics that is specifically Nazi, despite its obvious potential for developing some aspects of its Aryan racial agendas.[27] Race and class agendas dominated the concerns of most so-called 'progressives' of the 1920s and 1930s, who supported the practice of selective human breeding. For example, Marie Stopes (1880–1958), who founded the 'Society for Constructive Birth Control and Racial Progress' in 1921, aggressively advocated sterilisation (voluntary or forced) as a means of preventing the 'rotten and racially diseased' from endangering the 'higher and more beautiful forms of the human race'.[28] Programmes of artificial insemination were advocated in the Soviet Union in the 1920s as a means of social engineering; these were subsequently abandoned for ideological reasons.[29]

The appeal of eugenics to those inclined towards social engineering has never gone away, although it has long since lost any associations with progressive trends within Western culture, partly because of its associations with a discredited Nazi past, and partly because some of its core policies – such as forcible sterilisation – are alien to the fundamental values of a neo-liberal society. Nevertheless, it is not difficult to understand why some continue to find it attractive. If plants and animals can be bred selectively so that they are more resistant to disease, why not do the same with human beings? Selective breeding might easily eliminate those who would, in the eyes of some, be a 'burden' to society.[30] A social 'in-group' could manipulate the genetic future of its rival 'out-groups'.

The lure of the genetic reprogramming of humanity remains as powerful as ever. In his influential book *On Aggression* (1963), the ethologist Konrad Lorenz argued that while the human tendency to aggressiveness once served some important social roles, the advent of modern technology made this instinct fundamentally dangerous. It is no accident that Lorenz concluded his book by expressing the faint hope that human beings might one day find themselves subject to some form of genetic mutation that would turn them into affectionate creatures. For Lorenz, this would be a most welcome happenstance. But what if this were to be *made* to happen? What if we could take control of things?

Rebooting Humanity:
Transhumanism and the Enhancement of Humans

Classic forms of humanism, expressed in works such as Pico della Mirandola's *Oration on the Dignity of Man* (34–36), emphasise the beauty and elegance of human nature, taking delight in the complexity of the human body and the range of human achievements. More recently, however, schools of thought have emerged which see human nature as a 'work-in-progress, a half-baked beginning that we can learn to remould in desirable ways'.[31] If Milan Kundera is right in suggesting that our longing for paradise is actually a desire to escape from the limits of being human (4), might our dissatisfaction and restlessness cease if we could transcend our biological origins through technological enhancement? By becoming posthuman?

In the previous section, we noted the potential of selective breeding to 'improve' the human race – for example, by breeding out certain genetic deficiencies which make people more prone to disease. Yet there is another option for the transformation of humanity which avoids the social stigma that is now attached to eugenics – namely, the technological enhancement of humanity. The 'transhumanist' movement advocates 'fundamentally improving the human condition' through technology, thus allowing us to 'eliminate ageing and to greatly enhance human intellectual, physical, and psychological capacities'.[32] Through the creation and application of technology, human beings now have the capacity to transcend their biological limits. Yet awkward questions remain. Nick Bostrom, one of the most significant and influential transhumanist philosophers, rightly queries the simplistic equation of technological advances with the notion of progress itself. 'It may be tempting to refer to the expansion of technological capacities as "progress". But this term has evaluative connotations – of things getting better – and it is far from a conceptual truth that expansion of technological capabilities makes things go better.'[33]

What sort of technological enhancements might this mean? Here's one example.[34] It will one day be possible to develop artificial red blood cells, capable of transporting oxygen and carbon dioxide in human blood. These artificial cells would not be limited by the materials and pressures that arise naturally, and would thus be capable of performing far beyond the range of natural red blood cells.

Some will see this intervention in human development as 'playing at being God'. This is a fair point, although it needs careful nuancing. After all, human beings already depend extensively on scientific interventions – such as drugs and surgery – to promote the quality and extend the range of their lives. The question under discussion is whether transhumanism represents an extension of existing practices, or a new approach altogether, moving us into unfamiliar and disturbing ethical and social territory.

Here's another example. Transhumanists point out the importance of 'evolutionary lag' – the slow response of natural evolutionary processes to changes in our environment. Technology now allows us to insert certain genes for specific traits, thus bypassing evolutionary lag. A good example of this issue is lactose tolerance. While the development of lactose intolerance is adaptive for mammals since it makes weaning easier, increasing use of dairy products in human societies over the last five thousand years has given rise to a significant problem for many people. This development is so recent in evolutionary terms that genetic traits which deal with the problem have yet to diffuse to all human populations. So why not insert into the human genome a gene that has a desirable effect in this situation, or find some way of mimicking its effects?

Bostrom argues that such judicious technological enhancement of humanity might lead to the emergence of 'posthumans' with some or all of the following characteristics.

1. Individuals could expect to live for more than five hundred years.

2. A large section of the population would have cognitive capacities that are more than two standard deviations above the present human maximum.
3. Psychological suffering would become rare.[35]

There are obvious questions here. For example, is such a technologically enhanced human being actually something other than human? And are these technological enhancements capable of being transmitted genetically? If so, are we really talking about the emergence of a new species? Will 'posthumans' displace humans?

Transhumanists often assume that technological augmentation of natural human cognitive capacities will lead to moral excellence, in effect regarding selfishness and our innate tendencies towards destructive patterns of thought and action as due to mental retardation that will be remedied by boosting our cognitive capacities. It's an interesting suggestion. Yet its status is that of an unevidenced belief, a charming aspiration, which may lack any grounding in reality. Victor Ferkiss expressed a deep-seated concern a generation ago that cannot be ignored, given the pace of technological development. 'What if the new man combines the animal irrationality of primitive man with the calculated greed and power-lust of industrial man, while possessing the God-like powers granted him by technology? This would be the ultimate horror.'[36]

There are many issues to be debated, some of which are scientific. The most important is the 'Weissman barrier' – the basic principle that certain characteristics that are acquired by organisms in their lifetime cannot be transmitted genetically. For example, a woman who loses a hand in an accident does not produce children with only one hand, just as a man who develops his muscles through rigorous exercise regimes does not necessarily produce muscular offspring. At what point would a technologically enhanced humanity be able to transmit its enhancements genetically – if at all?

Yet most of the questions that arise do not concern how this could be done, but whether it should be done. Four concerns seem particularly significant here.

1. Technological enhancements are expensive, and their implementation will therefore merely increase global inequality. Those with the financial means will be able to extend their lifespans; everyone else has to continue as normal. The first posthumans may well turn out to be wealthy Americans.

2. A significant expansion of human lifespan immediately raises the concern noted by Thomas Malthus in his *Essay on the Principle of Population* (1798). Given that the earth has limited resources, it can only sustain a certain number of people. Although Malthus could not have foreseen the development of chemical fertilisers and genetically modified crops, giving enhanced yields, his point remains valid. Earth's capacity to produce food limits the size of the population. If human life expectancy rises to 500 years, the overall population must be reduced. Otherwise, the harsh control mechanisms so grimly noted by Malthus – war and famine – would kick in.

3. The assumption that there is a direct correlation between cognitive capacity and moral discernment is contestable. What if technological enhancement merely assists and enables humanity's seemingly inescapable and utterly irrational tendency to debase and destroy its own best achievements? To use more theological language, does the rise of transhumanism offer an escape from sin? Or does it make us even more vulnerable to it?

4. Will transhumanists end up treating humans as an inferior race? As Stephen Hawking points out, there is no compelling reason for believing that a technologically enhanced humanity will retain the same values and rationality as its originals.

None of these questions are easy to answer, yet they all need to be asked. It is far from clear that technological advance will lead us to wiser and better decisions than we have made in the recent past. Perhaps that helps us understand why some are suggesting that any possible enhancement of human *technological* capacity means that we also need a corresponding *moral* enhancement, if we are to cope with the new challenges that we will inevitably face.[37] But who will reprogramme us? After all, these 'moral enhancements' will be developed and chosen by morally questionable human beings, who could easily adapt them to advance their own vested interests and concerns.

We might be on the threshold of a terrifying Orwellian world in which powerful groups reprogramme us to replicate their beliefs and values, making critique of these impossible through a redirection of human rational processes. Aldous Huxley's *Brave New World* (1932) is a troubling account of such a development, in which human beings were reprogrammed in 'hatcheries and conditioning centres' to conform to the will of their controllers, including the generation of people of lower intelligence in the 'Social Predestination Room', to facilitate their conformity to the needs of the World State. Instead of having to think for themselves, they would simply conform to the group-think of the State.

So would such moral rebooting actually change things? The sad truth is that humanity has messed up this world more than any other animal. We have managed to change global climates, with unpredictable and potentially damaging long-term results; we have developed nuclear and pathological weapons capable of wiping out entire populations; and we have conducted campaigns of genocide unparalleled elsewhere in the animal kingdom.

So can we be hopeful for the future of the human race? Given that 99.9 per cent of all species that ever existed on earth are now extinct, what are the chances of our survival? There seems to be an emerging consensus that there is a serious risk that humanity's journey will come to a premature end – even if that consensus does not extend to any precise timeframe or probability judgement.[38] We could be wiped out in a natural extinction event; we could wipe ourselves out; or we might be transformed into another species – but perhaps this time, as transhumanism suggests, by taking charge of our own evolutionary future. We just don't know. And we have to live in the light of this lack of knowledge. But maybe that's better than believing that the future is merely some kind of fantasy projection of the present. An uncritical belief in the uninterrupted upward trajectory of human progress can lead us into illusory and dystopian worlds, whose plausibility is sustained largely by the suppression of evidence and reflection.

Yet there is another point that needs to be made here. Earlier, we noted Einstein's problem with the question of the 'Now'. Viewing human life solely through the interpretative lens of relativity theory could not help him – or anyone – grasp the existential importance of the present moment in human experience. As we saw earlier in our reflections on meaning, human beings naturally consider this to be important. The same problem emerges with ideologies which focus on the theme of progress. Why? Terry Eagleton puts his finger on the point at issue: 'For the ideology of progress . . . all moments are devalued by the fact that each of them is no more than a stepping-stone to a successor, the present a mere gangplank to the future. Every point of time is diminished in relation to an infinity of points still to come.'[39]

This ideology of progress holds that some point in the indeterminate future is thus held out as possessing value and significance; the present moment only has significance in that it contributes to this distant and invisible goal. The present is thus reduced to little more than an existential vacuum, empty of any intrinsic significance save as a stepping-stone to an imagined future that may turn out to be nothing more than an illusory promissory note. If such a controlling and limiting ideology allows us to have hope, it is for the transformation of the future. Yet human intellectual history is littered with the wrecks of aspirations that proved to be groundless, misguided or simply insane. Who knows if today's cherished dreams will turn out to be tomorrow's discarded illusions?

The Hope of Immortality: Technology or Theology?

John Gray had little doubt about one of the core human obses-
sions. 'Longing for everlasting life, humans show that they remain
the death-defined animal.'[40] Yet even those who do not share this
fixation on human mortality are obliged to recognise that there
are limits placed on our lifespan. We may seek to extend our lives,
yet this will simply add further to the stress on our planet's limited
resources, and make war or famine even more likely. So why this
anxiety in the first place?

Many answers have been given. The sociologist Peter Berger
suggested that human mortality was the ultimate unbearable
truth, pointing to the terrifying meaninglessness and chaos
which characterised human existence in this world. Society
tried to defend and protect people from this unbearable knowledge, shielding them against this terror by asserting the
existence of meaning and order in a seemingly meaningless and
chaotic universe.[41] Human beings need to be protected by illusions of meaning against this meaningless, chaotic realm of
disorder, disintegration and death. J.R.R. Tolkien spoke of
human beings spinning 'wish-fulfilment dreams' to console and
cheat 'our timid hearts', while seeing this human instinct as a
sign of some greater horizon that called out to be explored.[42]
This human reluctance to accept our own mortality also stood
at the heart of Ernest Becker's Pulitzer Prize-winning mono-
graph *The Denial of Death* (1973). For Becker, human beings
are driven by the desire to deny death, transcend it, or create
meaning in its face.

History is generous in its benevolent provision of amusing
examples to illustrate human attempts to deny, cheat or conquer
death. One of the most entertaining is the attempt made by the
Soviet 'Immortalization Commission' to preserve Lenin's body
after his death in 1924 – not merely as a potent symbol of the
Russian Revolution, but as an assertion of the human capacity to
achieve immortality under Marxism-Leninism.[43] Leonid Krasin

was an engineer who believed that the dead could be technologically resurrected. Who more than Lenin deserved to be the first to share this privilege? In a series of remarkable experiments, Krasin used refrigeration to preserve Lenin's body, not merely so that its public display might edify the Soviet masses, but as a demonstration of the Soviet capacity to vanquish death. After all, the Soviet Union arose through the forceful overthrow of imperial Russia. Why not end the tyranny of death in the same way?

Sadly, Krasin's primitive cryogenic experiments were doomed to failure. Lenin's body soon began to show unmistakeable signs of decay. Krasin demanded that a better refrigerator should be imported from Germany to stop this process of deterioration. Yet it continued unabated and irresistibly. In his death, Lenin demonstrated his true humanity. His nose began to lose its shape, and his eyes began to recede into their sockets. In the end, the only solution was to embalm Lenin's body regularly, and hope nobody would notice that death had neither been defeated nor transfigured.

Yet there was a clear ideological motivation for this procedure. Marxism-Leninism proclaimed itself as having displaced an outdated and outmoded Christianity. This, unfortunately, had been achieved largely through force of arms, rather than force of argument. Churches were closed or destroyed, often by dynamiting; priests were imprisoned, exiled or executed. On the eve of the Second World War there were only 6,376 clergy remaining in the Russian Orthodox Church, compared with the pre-revolutionary figure of 66,140. In 1917, there were 39,530 churches in Russia; by 1940, only 950 remained functional.[44]

A key player in this atheist crusade was the League of Militant Atheists, a semi-official coalition of various political forces which operated within the Soviet Union from 1925 to 1947.[45] With the slogan, 'The Struggle against Religion is a Struggle for Socialism', the group set out to destroy the credibility of religion through social, cultural and intellectual manipulation. Its carefully orchestrated campaigns involved using newspapers, journals, lectures and films to persuade Soviet citizens that religious beliefs and practices were irrational and destructive. Good Soviet citizens, they declared, ought to embrace a scientific, atheistic worldview. The good citizens of the Soviet Union nevertheless clung obstinately to their Christian faith – not as a mark of intellectual backwardness, but because they preferred to trust this narrative rather than its more recent and untested rival, which already seemed to them to be in trouble.

Christianity proclaimed the resurrection of the body, and offered the hope of eternal life. Marxism-Leninism thought that it could do better than that: it could overcome death itself, and offer a secular hope. The advance of knowledge within a technocratic Soviet Union would make it possible for humanity to conquer death without the need for divine assistance. Although most Soviet ideologists were emphatic that Marxism-Leninism was not a religion, its vision for human immortality certain made it seem so to its secular admirers. The problem, of course, was that it failed to deliver on this core promise. Transhumanism has taken up this challenge, with a promise to extend human life. Yet there is no talk of immortality. Wisely, transhumanism has limited itself to the postponement of the inevitable, rather than promising the impossible.

There is much more that needs to be explored in this book. Yet it is time to bring these brief reflections on human beings to an end.

13

Endings: Some Brief Musings

'My life seemed to be passing before me, not in a flash as
it is said to do for those about to drown, but in a sort of
leisurely convulsion, emptying itself of its secrets and its
quotidian mysteries in preparation for the moment when
I must step into the black boat on the shadowed river with
the coin of passage cold in my already coldening hand.'[1]

John Banville

This work is bookended by two quotes from *The Sea*, a masterly
work by the Irish writer John Banville (born 1945). In the above
passage, rich with symbolic associations, Banville holds before us
the chilling vision of stepping into Charon's boat to be ferried
across the Stygian darkness to the place of the dead without
having managed to figure out what life was all about in the first
place. We leave behind a mysterious and puzzling world, through
which we passed uncomprehendingly, and prepare to enter
another realm about which we know even less. It is a thought that
many prefer to ignore.

Life is a mystery – something with so many impenetrable and
uncomprehended dimensions that our minds simply cannot take
it in. We can only cope with such a mystery by filtering out what
little we can grasp, thus reducing it to what our minds can accommodate, so that reality is displaced by what is rationally
manageable. Some withdraw from any intellectual engagement of
this kind, believing that it is pointless. The wise persist, however,
knowing that any worthwhile insights come only by an uncomfortable but necessary stretching of our minds.

For Richard Dawkins, our universe is genuinely mysterious.
How can the human mind, seemingly designed to cope with the

quotidian threats and opportunities upon which survival depends, conceivably take in the vastness of our universe? 'Modern physics teaches us that there is more to truth than meets the eye; or than meets the all too limited human mind, evolved as it was to cope with medium-sized objects moving at medium speeds through medium distances in Africa.'[2] I agree with him entirely. I fear that some of his more dogmatic rationalist friends might demur at his reference to an 'all too limited human mind', yet Dawkins has science on his side at this point, and they will just have to get used to this. Recognition of the limitations of the human mind is not, however, an insight that is restricted to modern physics, but is a signature characteristic of the human predicament.

Some suggest that mystery is just a superstitious person's way of referring to an irrationality. As a slogan, this suggestion is both slick and simple; as a guide to reality, however, it is superficial and deeply misleading. Those of us who have studied quantum theory know it has developed its own rationality of our fuzzy world, which calls into question inadequate common-sense conceptions of what is reasonable, shaped by our limiting experience of reality. Lesser rationalities – such as those of the bygone 'Age of Reason' – tend to evade the challenge of more expansive visions of rationality by dubbing them 'irrational'. In fact, their views are corrected and relativised by scientific advance, both in the natural and social sciences. Karl Marx, Charles Darwin, Sigmund Freud and Albert Einstein have all played critical roles in challenging such naïve visions of rationality.

More recently, postmodernism has claimed that science, religion and philosophy are tainted with cultural influences, causing us to conflate truth with power. There is a legitimate debate here about the extent to which this is indeed so, yet the evidence suggests there is a genuine case to answer and a legitimate cause for concern. There can be little doubt that further chastening of our inflated ideas of rationality awaits us. The question that remains on the table is this: what can be learned from recognising that the human mind is 'all too limited'?

As I bring this work to a close, I see three points that have clearly emerged from the chastening of human rationality that we have considered in this work. None are religious in themselves, yet all are religiously significant. They all call out for further thought.

1. Humility: Reality is a Lot Bigger than We Are

The proper attitude to the universe is humility – a respectful appreciation of its spatial and temporal vastness, in the face of which we seem insignificant. We can only have a partial and incomplete grasp of our universe. This does not lead to a relativist anarchy, in which all views are held to be equally good. Our beliefs about the universe need to be warranted or rationally motivated – that is to say, there must be reasons for believing that they might be right. Yet as the progress of science itself over the last few centuries makes clear, what one generation regarded as correct is discarded by another generation as false or inadequate.

In a powerful visual image, Paul of Tarsus speaks of the limits of our vision of reality. We now only 'see in a mirror, dimly' (1 Corinthians 13:12). To be human is to see only part of a 'big picture' of reality, and to learn to live with a deep sense of frustration that we cannot see more, mingled with some slight satisfaction that we can see anything at all. The enemy here is not faith, but arrogance – the conceited belief that our limited view is right, and for that reason others are wrong.

2. Generosity: We Will Have to Learn to Live With Unresolved Questions

Humans long for certainty and clarity. While the Enlightenment's quest for clear and certain ideas proved impossible to attain, save in very limited contexts, it was nevertheless a very human aspiration. That's the sort of knowledge that we yearn for. But save in the limited realms of logic and mathematics, it's not what we can realistically expect to achieve. Aspirations to Cartesian clarity have faded, to be replaced with postmodern hesitation. It has simply not proved possible to rescue the Enlightenment's rational ideals from the sustained criticisms of Martin Heidegger, the later Wittgenstein, and Hans-Georg Gadamer. Even in the natural sciences, we have to accept the provisionality of scientific theories. Our beliefs may be warranted; they might not, however, be right. 'Neither science nor religion can entertain the hope of establishing logically coercive proof of the kind that only a fool could deny.'[3]

This means that we need to learn to live with unresolved questions, hoping to find answers that may be trusted, yet so often cannot be proved to be true. I come back to Bertrand Russell's insight (92) that our condition is such that we need to cope with living without the hope of certainty.[4] It is not difficult to point to ultimate questions of life that remain frustratingly resistant to definitive answers. What, for example, lies beyond death? Maybe death ends everything. Maybe we must step into Charon's boat for our final journey to a dark underworld. Maybe we might enter a New Jerusalem, a place of safety and rejoicing. Nobody knows for certain, although I know what I believe and hope – and why. But that doesn't mean that I regard others who disagree with me as deluded fools. I gladly leave that sort of rhetoric to cultural fundamentalists.

For perhaps the most important point to appreciate from such a line of thought is this. Given that life's great questions remain tantalisingly open, we need to be generous towards those whose answers do not coincide with our own. Sir Isaiah Berlin famously argued that ultimate human values – such as questions of justice or the God-question – necessarily lead to conflict. They cannot be reconciled or resolved by rational calculation, since they cannot be measured. Nor, in Berlin's view, could they be sorted out by philosophy, which Berlin was inclined to think incapable of solving anything that is of ultimate importance to us as human beings. The divergence of opinions and values in human thought is not ultimately a mark of mental fragility or malfunction, but of the intractable nature of the world within which we live and think.[5] If Berlin is right, human beings end up with 'a diversity of narratives about themselves, none of which has the authority of a metanarrative'.[6] And since reality is too complex to compel us to agree on its right interpretation, we might at least be civil to each other, as we try to make sense of things as best we can.

3. Wonder: A Willingness to Expand our Vision

Our sense of wonder at the solemn stillness of a starlit night, or the crystal purity of a northern landscape, elicits more than an experience of beauty; it discloses the limited capacity of our minds, and invites us to expand our mental vision to embrace the universe as it actually is, rather than to try and reduce it to what we find convenient or manageable. We face a constant struggle within and against ourselves, as we stare down the temptation to desiccate and dissect the mystery of ourselves and our universe. Wordsworth's words still ring true:

> Our meddling intellect
> Mis-shapes the beauteous forms of things:–
> We murder to dissect.[7]

A 'mystery', in the theological sense of that term, is something that ultimately triumphs over the human instinct to reduce and simplify, exposing the limits of the systems and frameworks within which we try to trap reality.

Recent psychological studies of the human experience of awe have highlighted the tension that arises from wanting to *understand* what we experience, yet simultaneously wanting to *sustain* the experience itself.[8] At their best, both science and religion try to *preserve* this mystery, affirming the intelligibility of our universe without reducing this to what the 'all too limited human mind' can accommodate. In Christianity, this is evident in the

creative tension between theology and worship, which paradoxically celebrates both the fact that so much of God can be grasped,
however inadequately, by the human mind (and hence leads to
theology), while at the same time so much remains beyond the
human capacity to understand (and hence leads to worship, in the
sense of acknowledging that the greatness and majesty of God
eludes verbal analysis, and is best expressed in praise and
adoration).

Conclusion

This book has reflected on the great mystery of human identity,
and especially our attempts to make sense of ourselves and our
world as we journey on the Road (51–55, 97). The proper human
response to that mystery is a sense of humility in the presence of
something greater than us; an intellectual generosity, not a narrow
dogmatism, in trying to make sense of it; and perhaps above all, a
feeling of awed wonder that we are temporarily placed within
such a vast and beautiful universe, and are capable of reflecting on
its deeper significance – as well as our own.

It is easy to be drawn to ideologues who try to short-circuit or
bypass the problem of our 'all too limited human mind', offering
us the cosy and smug certainties of in-groups or the strident
slogans of fundamentalists. These imagined certainties demand a
God's-eye view of reality, to which neither they nor we have access.
Our place is on the Road, not on the Balcony – not a settled habitation of detached privilege, but a process of journeying in hope
through an opaque and puzzling world.

The approach to the mystery of human identity and meaning
set out in this book is both informed by science and nourished
and enriched by the Christian tradition. We indeed are 'meaning-
seeking animals'. Just as we seek food in order that we may
survive, so we seek meaning in order that we may flourish. Perhaps
we are, as John Banville suggests, a 'little vessel of strangeness'
sailing through the 'muffled silence' of an 'autumn dark'. Perhaps

we have been forced to pause in the middle of our lives, to use Dante's famous image, as we realise that we are lost in a dark wood, unsure either of where we are or where we are meant to be.[9]

Dante's *Divine Comedy* is not the only theologically informed account of our quest for meaning, nor the only substantial work of literature to be organised around the image of a journey through life. Yet many regard it as the most acutely observed and finely expressed account of our search for significance. Perhaps in closing, we might reflect on what one of Dante's leading interpreters saw as its core vision.

> We find affirmed with the utmost clarity and consistency the fundamental Christian proposition that the journey to God is the journey into reality. To know all things in God is to know them as they really are, for God is the only absolute and unconditioned Reality, of whose being all contingent realities are at best the types and mirror, at worst the shadows and distortions.[10]

We need a vision such as this to sustain us as we pass through this world of dark shadows and pale reflections, thinking and wondering as we journey. We are unable to stand above the Road and see the 'big picture' for ourselves – but we can trust that there is one, giving us meaning and purpose as we travel on that Road.

Acknowledgements

This book is an expanded version of my opening keynote lecture at the 2015 conference of the Ian Ramsey Centre for Science and Religion at Oxford University. The conference brought together experts from the fields of anthropology, biology, philosophy, psychology and theology to present and debate the implications of recent research on the 'human difference' – the question of what distinguishes human beings from other living species, and what conceptual frameworks might be helpful in clarifying the nature of this difference. The rich discussion that followed my lecture stimulated me to write it up as a book. In expanding this lecture, I was able to engage some of the probing questions that arose during that extended discussion at the Ian Ramsey Centre. I owe much to those who have helped me develop these ideas in the last few years. While I can never hope to acknowledge my full debt to others, I can at least thank those who have proved to be such a stimulus to my own thinking in this area, especially Joanna Collicutt, Mary Midgley, Andrew Pinsent, Donovan Schaefer, Raymond Tallis, Graham Ward and Johannes Zachhuber.

Notes

Chapter 1

1. John Banville, *The Sea*, New York: Vintage Books, 2005, 53.
2. For an excellent study of this phenomenon from 1933 to 1973, see Mark Greif, *The Age of the Crisis of Man: Thought and Fiction in America, 1933–1973*, Princeton, NJ: Princeton University Press, 2015.
3. Reinhold Niebuhr, *The Nature and Destiny of Man: A Christian Interpretation*, 2 vols, London: Nisbet, 1941–3, vol. 2, 214.
4. Milan Kundera, *The Unbearable Lightness of Being*, London: Faber & Faber, 1995, 287–90.
5. Robert C. Fuller, *Wonder: From Emotion to Spirituality*, Chapel Hill, NC: University of North Carolina Press, 2006, 54–68.
6. See Pierre Hadot, *What Is Ancient Philosophy?*, Cambridge, MA: Harvard University Press, 2002. Note especially his reflections on Marcus Aurelius, who was 'trying to do what, in the last analysis, we are all trying to do: to live in complete consciousness and lucidity, to give to each of our instants its full intensity, and to give meaning to our entire life'. Pierre Hadot, *The Inner Citadel: The Meditations of Marcus Aurelius*, Cambridge, MA: Harvard University Press, 1998, 312–13.
7. Susan R. Wolf, 'The Meanings of Lives', in *The Variety of Values: Essays on Morality, Meaning, and Love*, New York: Oxford University Press, 2015, 89–106.
8. Henry D. Thoreau, *Walden,* New York: Thomas Crowell, 1910, 17. For a philosopher's reflections on this concern, see Pierre Hadot, 'Il y a de nos jours des professeurs de philosophie, mais pas de philosophes'. In *Exercices spirituels et philosophie antique*. Paris: A. Michel, 2002, 333–42.
9. Kundera, *The Unbearable Lightness of Being*, 135.
10. Happily, some professional philosophers are more than willing to engage with such questions. See especially the highly engaging study of Rebecca Goldstein, *Plato at the Googleplex: Why Philosophy Won't Go Away*, New York: Pantheon, 2014.

11. See, for example, Paul T.P. Wong, *The Human Quest for Meaning: Theories, Research, and Applications*, 2nd edn. New York: Routledge, 2012.

12. See, for example, Karen Gasper and Gerald L. Clore, 'Attending to the Big Picture: Mood and Global Versus Local Processing of Visual Information', *Psychological Science* 13 (2002): 34–40; Joshua A. Hicks and Laura A. King, 'Meaning in Life and Seeing the Big Picture: Positive Affect and Global Focus'. *Cognition and Emotion* 21, no. 7 (2007): 1577–84.

13. G.K. Chesterton, *Autobiography*, San Francisco: Ignatius, 2006, 99.

14. See Jean-Charles Falardeau, 'Le sens du merveilleux', in *Le merveilleux: Deuxième colloque sur les religions populaires*, ed. Fernand Dumont, Jean-Paul Montminy and Michel Stein. Québec: Presses de l'Université Laval, 1973, 143–56.

15. Hermann Hesse, *Mit dem Erstaunen fängt es an: Herkunft und Heimat, Natur und Kunst*, Frankfurt: Suhrkamp Verlag, 1986.

16. A point argued with particular force and clarity in Roger Wagner and Andrew Briggs, *The Penultimate Curiosity: How Science Swims in the Slipstream of Ultimate Questions*, Oxford: Oxford University Press, 2016.

17. Mary-Jane Rubenstein, *Strange Wonder: The Closure of Metaphysics and the Opening of Awe*, New York: Columbia University Press, 2010, 1–11.

18. For our different positions, see Richard Dawkins, *An Appetite for Wonder: The Making of a Scientist*, London: Bantam Press, 2013; Alister E. McGrath, *Inventing the Universe: Why We Can't Stop Talking About Science, Faith and God*, London: Hodder & Stoughton, 2015.

19. For this phrase, see Richard Dawkins, *A Devil's Chaplain: Selected Essays*, London: Weidenfeld & Nicolson, 2003, 19.

20. Aristotle, *Metaphysics*, 982b. See also Plato, *Theaetetus*, 154b–155c. For useful reflections on this theme, see Jerome Miller, *In the Throe of Wonder*. Albany: State University of New York Press, 1992, 11–52.

21. Michael Polanyi, 'Science and Reality', *British Journal for the Philosophy of Science* 18, no. 3 (1967): 177–96, especially 177–9.

22. http://www.nytimes.com/1990/05/15/books/once-more-admired-than-bought-a-writer-finally-basks-in-success.html

23. Rüdiger Imhof, *John Banville: A Critical Introduction*, Dublin: Wolfhound Press, 1989, 62. See also Elke D'Hoker, *Visions of Alterity:*

Representation in the Works of John Banville. Amsterdam: Rodopi, 2004, 17–48.

24. Lewis's own views on this problem should be noted here: see Alister E. McGrath, 'An Enhanced Vision of Rationality: C. S. Lewis on the Reasonableness of Christian Faith'. *Theology* 116, no. 6 (2013): 410–17.

25. Albert Einstein, *Ideas and Opinions*. New York: Crown Publishers, 1954, 41–9.

26. Peter B. Medawar and Jean Medawar, *The Life Science: Current Ideas of Biology*. London: Wildwood House, 1977, 171.

27. A good starting point is Michael J. MacKenzie and Roy F. Baumeister, 'Meaning in Life: Nature, Needs, and Myth'. In *Meaning in Positive and Existential Psychology*, ed. Alexander Batthyany and Pninit Russo-Netze. New York: Springer, 2014, 25–38.

28. Massimo Pigliucci, 'New Atheism and the Scientistic Turn in the Atheism Movement'. *Midwest Studies in Philosophy* 37, no. 1 (2013): 142–53; quote at p. 144. See further Michael D. Aeschliman, *The Restitution of Man: C. S. Lewis and the Case against Scientism*. Grand Rapids, MI: Eerdmans, 1998; Richard N. Williams and Daniel N. Robinson, eds. *Scientism: The New Orthodoxy*. London: Bloomsbury, 2015.

29. Bernard Williams, *Philosophy as a Humanistic Discipline*. Princeton, NJ: Princeton University Press, 2005, 182. Williams's critique of this tendency is highly perceptive, and worth careful study.

30. William James. *The Will to Believe*. New York: Dover Publications, 1956, 51.

31. See Lloyd P. Gerson, *God and Greek Philosophy: Studies in the Early History of Natural Theology*. London: Routledge, 1994.

32. Susan R. Wolf, *Meaning in Life*. Princeton, NJ: Princeton University Press, 2010, 10–11.

33. Alexander Wood, *In Pursuit of Truth: A Comparative Study in Science and Religion*. London: Student Christian Movement, 1927, 102.

34. Salman Rushdie, *Is Nothing Sacred?* The Herbert Read Memorial Lecture. Cambridge: Granta, 1990, 8–9.

35. Sam Harris, *Waking Up: Searching for Spirituality Without Religion*. London: Transworld Publishers, 2014.

36. Victoria Harrison, 'The Pragmatics of Defining Religion in a Multi-Cultural World'. *International Journal for Philosophy of Religion* 59 (2006): 133–52.

37. For evaluations of this distorted idea, see Jonathan Haidt, *The Righteous Mind: Why Good People Are Divided by Politics and Religion*. New York: Pantheon Books, 2012. The 'New Atheism' tends to assume that American Protestantism provides a template for religion as a whole: see Donovan Schaefer, 'Blessed, Precious Mistakes: Deconstruction, Evolution, and New Atheism in America'. *International Journal for Philosophy of Religion* 76 (2014): 75–94.

38. Keith Yandell, *Philosophy of Religion: A Contemporary Introduction*. London: Routledge, 1999, 16.

39. On this 'universal Darwinism', see Richard Dawkins, 'Darwin Triumphant: Darwinism as Universal Truth'. In *A Devil's Chaplain: Selected Essays*. London: Weidenfeld & Nicolson, 2003, 78–90.

40. Mary Midgley, *Evolution as a Religion: Strange Hopes and Stranger Fears*. 2nd edn. London: Routledge, 2002, 17–18.

41. Edward O. Wilson, *Consilience: The Unity of Knowledge*. New York: Vintage, 1999, 294.

42. For critical assessments of Wilson's approach, see Ullica Segerstrale, 'Wilson and the Unification of Science'. *Annals of the New York Academy of Sciences*, no. 1093 (2006): 46–73; Abraham G. Gibson, 'Edward O. Wilson and the Organicist Tradition'. *Journal of the History of Biology* 46 (2013): 599–630.

43. Alister E. McGrath, *Inventing the Universe: Why We Can't Stop Talking about Science, Faith and God*. London: Hodder & Stoughton, 2015.

Chapter 2

1. Raymond Carver, 'Late Fragment'. In *All of Us: The Collected Poems*. London: Harvill Press, 1996, 294.

2. For an excellent survey, see Louis Tay and Ed Diener, 'Needs and Subjective Well-Being around the World'. *Journal of Personality and Social Psychology* 101, no. 2 (2011): 354–65.

3. Sigmund Freud, 'One of the Difficulties of Psycho-Analysis'. *Journal of Mental Science* 67 (1921): 34–9.

4. Gabriel W. Finkelstein, *Emil du Bois-Reymond: Neuroscience, Self, and Society in Nineteenth-Century Germany*. Cambridge, MA: MIT Press, 2013.

5. Dennis Danielson, '[The Myth] that Copernicanism demoted Humans from the Center of the Cosmos'. In *Galileo Goes to Jail and Other*

Myths about Science and Religion, ed. Ronald L. Numbers. Cambridge, MA: Harvard University Press, 2009, 50–8.

6. Iris Murdoch, *The Sovereignty of Good*. London: Routledge, 2001, 82.

7. Raymond Tallis, *Aping Mankind: Neuromania, Darwinitis and the Misrepresentation of Humanity*. London: Routledge, 2014, 349.

8. Robin Dunbar, *The Human Story*. Faber & Faber, 2004, 197–9.

9. For a good overview, see Richard N. Williams and Daniel N. Robinson, eds, *Scientism: The New Orthodoxy*. London: Bloomsbury, 2015.

10. Edward O. Wilson, *The Meaning of Human Existence*. New York: W.W. Norton, 2014, 161.

11. Stephen Hawking and Leonard Mlodinow, *The Grand Design*. New York: Bantam Books, 2010, 5.

12. For a good discussion, see Paul Thagard, 'Why Cognitive Science Needs Philosophy and Vice Versa'. *Topics in Cognitive Science* 1 (2009): 237–54.

13. Charles A. Coulson, *Christianity in an Age of Science*. London: Oxford University Press, 1953, 21.

14. Coulson, *Christianity in an Age of Science*, 20. See further Alister E. McGrath, 'Multiple Perspectives, Levels, and Narratives: Three Models for Correlating Science and Religion'. In Louise Hickman and Neil Spurway, eds, *Forty Years of Science and Religion*. Newcastle: Cambridge Scholars, 2016, 10–29.

15. Coulson, *Christianity in an Age of Science*, 21.

16. Coulson, *Christianity in an Age of Science*, 21.

17. Francis Crick, *The Astonishing Hypothesis: The Scientific Search for the Soul*. London: Simon & Schuster, 1994, 3; 11.

18. For a detailed discussion of Dawkins' views, see Alister E. McGrath, *Dawkins' God: From* The Selfish Gene *to* The God Delusion. 2nd edn. Oxford: Wiley-Blackwell, 2014.

19. Richard Dawkins, *River Out of Eden: A Darwinian View of Life*. London: Weidenfeld & Nicholson, 1995, 133.

20. Richard Dawkins, *The Selfish Gene*. 2nd edn. Oxford: Oxford University Press, 1989, 21.

21. Noted in John Dupré, *Human Nature and the Limits of Science*. Oxford: Clarendon Press, 2001, 1–3.

22. Cited in Michael Plekon, ed., *Tradition Alive: On the Church and the Christian Life in Our Time*. Oxford: Sheed & Ward, 2003, 172.

23. For my own use of anthropology, see Alister E. McGrath, 'Narratives of Significance: Reflections on the Engagement of Anthropology

and Christian Theology'. In *A Theologically Engaged Anthropology*, ed. Derrick Lemons, New York: Oxford University Press, forthcoming.

24. See, for example, John B. Vickery, *The Literary Impact of The Golden Bough*. Princeton, NJ: Princeton University Press, 2015. For concerns about its approach, see Jonathan Z. Smith, 'When the Bough Breaks'. *History of Religions* 12, no. 4 (May, 1973): 342–71.

25. Derek Freeman, *Margaret Mead and Samoa: The Making and Unmaking of an Anthropological Myth*. Cambridge, MA: Harvard University Press, 1983.

26. David I. Kertzer, 'Social Anthropology and Social Science History'. *Social Science History* 33, no. 1 (2009): 1–16.

27. Rodolfo Maggio, 'The Anthropology of Storytelling and the Storytelling of Anthropology'. *Journal of Comparative Research in Anthropology and Sociology* 5, no. 2 (2014): 89–106.

28. See further the detailed discussion in McGrath, 'Multiple Perspectives, Levels, and Narratives' (ch. 2, note 14 above).

Chapter 3

1. Augustine of Hippo, *Confessions*, X.vii.15.
2. Immanuel Kant, *Gesammelte Schriften*, 30 vols, Berlin: Reimer, 1902, vol. 5, 161.
3. Alexander Pope, *Essay on Man*, II.1–2.
4. Pope, *Essay on Man*, II.17–18.
5. For discussion of this general question around this time, see Kevin L. Cope, *Criteria of Certainty: Truth and Judgment in the English Enlightenment*. Lexington, KY: University Press of Kentucky, 1990.
6. Pope, *Essay on Man*, II.19–30; 35–8.
7. For example, see James Gilligan, *Violence: Our Deadly Epidemic and Its Causes*. New York: Vintage Books, 1997.
8. For a good discussion, see Rowan Williams, *The Edge of Words: God and the Habits of Language*. London: Bloomsbury, 2014.
9. Rowan Williams, *On Augustine*. London: Bloomsbury, 2016, 22.
10. Liliann Manning, Daniel Cassel and Jean-Christophe Cassel, 'St. Augustine's Reflections on Memory and Time and the Current Concept of Subjective Time in Mental Time Travel'. *Behavioral Sciences* 3, no. 2 (2013): 232–43.

11. See, for example, Karl K. Szpunar, 'On Subjective Time'. *Cortex* 47 (2011): 409–11; idem, 'Evidence for an Implicit Influence of Memory on Future Thinking'. *Memory & Cognition* 38 (2010): 531–40.

12. See especially Alister E. McGrath, 'The Enigma of Autobiography: Critical Reflections on *Surprised by Joy*'. In *The Intellectual World of C. S. Lewis*. Oxford: Wiley-Blackwell, 2013, 7–30.

13. Lydia Schumacher, *Divine Illumination: The History and Future of Augustine's Theory of Knowledge*. Malden, MA: Wiley-Blackwell, 2011.

14. See the detailed discussion in Jesse Couenhoven, *Stricken by Sin, Cured by Christ: Agency, Necessity, and Culpability in Augustinian Theology*. New York: Oxford University Press, 2013.

15. For example, see Charles G. Nauert, *Humanism and Renaissance Civilization*. Basingstoke: Ashgate, 2012.

16. The Latin title is 'Oratio de dignitate hominis'.

17. The rich diversity of Renaissance understandings of human nature is considered in Andreas Höfele and Stephan Laqué, eds, *Humankinds: The Renaissance and Its Anthropologies*. Berlin: Walter de Gruyter, 2011.

18. Giovanni Pico della Mirandola, *Oration on the Dignity of Man*. Chicago: Gateway Editions, 1956, 7 [I have altered the translation at points for accuracy].

19. For this translation, see William H. Race, *Pindar: Olympian Odes; Pythian Odes*. Cambridge, MA: Harvard University Press, 2002, 239.

20. For reflections on these themes, see Pierre Hadot, *What Is Ancient Philosophy?* Cambridge, MA: Harvard University Press, 2002.

21. G.K. Chesterton, *What's Wrong with the World*. San Francisco: Ignatius Press, 1994, 180.

22. Isaac Kalimi and Seth Francis Corning Richardson, eds, *Sennacherib at the Gates of Jerusalem: Story, History and Historiography*. Leiden: Brill, 2014. For an overstatement of these concerns, see Antti Laato, 'Assyrian Propaganda and the Falsification of History in the Royal Inscriptions of Sennacherib'. *Vetus Testamentum* 45, no. 3 (1995): 198–226.

23. Iris Murdoch, *The Sovereignty of Good*. London: Routledge & Kegan Paul, 1970, 82.

24. Murdoch, *Sovereignty of Good*, 91.

25. Murdoch, *Sovereignty of Good*, 50.

26. Iris Murdoch, *Existentialists and Mystics: Writings on Philosophy and Literature*. London: Chatto & Windus, 1997, 14.

27. Iris Murdoch, *Metaphysics as a Guide to Morals*. London: Penguin, 1992, 321.
28. Iris Murdoch, 'Ethics and the Imagination'. *Irish Theological Quarterly* 52 (1986): 81–95; quote at p. 90.
29. Iris Murdoch, *The Fire and the Sun: Why Plato Banished the Artists*. Oxford: Oxford University Press, 1977, 78.
30. Murdoch, *Sovereignty of Good*, 93.
31. For a fuller consideration of this theme, see Maria Antonaccio, *Picturing the Human: The Moral Thought of Iris Murdoch*. Oxford: Oxford University Press, 2000.
32. Murdoch, *Sovereignty of Good*, 58.
    Alan Grafen, 'The Intellectual Contribution of *The Selfish Gene* to Evolutionary Theory'. In *Richard Dawkins: How a Scientist Changed the Way We Think*, ed. Alan Grafen and Mark Ridley. Oxford: Oxford University Press, 2006, 66–74; quote at p. 73.
33. Typewritten text of 1966 reproduced in Richard Dawkins, *An Appetite for Wonder: The Making of a Scientist – A Memoir*. London: Bantam, 2013, 200.
34. Richard Dawkins, *The Selfish Gene*. 2nd edn. Oxford: Oxford University Press, 1989, 9–10.
    See Marek Kohm, 'To Rise Above'. In *Richard Dawkins: How a Scientist Changed the Way We Think*, ed. Alan Grafen and Mark Ridley. Oxford: Oxford University Press, 2006, 248–54.
35. Frans de Waal, *Our Inner Ape: The Best and Worst of Human Nature*. London: Granta Books, 2006, 21.
36. Dawkins, *The Selfish Gene*, 10.
37. Dawkins, *The Selfish Gene*, 200–1.

Chapter 4

1. T.S. Eliot, *The Dry Salvages*. London: Faber & Faber, 1941.
2. To understand his importance, readers might like to consider Carlo Brentari, *Jakob von Uexküll: The Discovery of the Umwelt between Biosemiotics and Theoretical Biology*. New York: Springer, 2015.
3. Jakob von Uexküll, *A Foray into the Worlds of Animals and Humans: With a Theory of Meaning*. Minneapolis: University of Minnesota Press, 2010, 53.
4. For a good discussion of this view, see Richard Rorty, *Objectivity*,

Relativism and Truth. Philosophical Papers. Cambridge: Cambridge University Press, 1991, 21–34.

5. Richard Rorty, *Consequences of Pragmatism*. Minneapolis, MN: University of Minneapolis Press, 1982, 166.

6. For comment on this complex work, see Harry M. Solomon, *The Rape of the Text: Reading and Misreading Pope's 'Essay on Man'*. London: University of Alabama Press, 1993.

7. Ludwig Wittgenstein, *Notebooks, 1914–1916*. New York: Harper, 1961, 75.

8. John V. Taylor, *The Go-Between God*. London: SCM Press, 1979, 19.

9. Taylor, *The Go-Between God*, 19.

10. C.S. Lewis, *Essay Collection*. London: HarperCollins, 2002, 21.

11. For a discussion of this point, see Stephen M. Barr, *Modern Physics and Ancient Faith*. Notre Dame, IN: University of Notre Dame Press, 2003, 188.

12. C.S. Lewis, *Surprised by Joy*. London: HarperCollins, 2002, 197. For a full account of Lewis's conversion, see Alister E. McGrath, *C. S. Lewis: A Life – Eccentric Genius, Reluctant Prophet*. London: Hodder & Stoughton, 2013, 131–51.

13. C.S. Lewis, *Studies in Medieval and Renaissance Literature*. Cambridge: Cambridge University Press, 2007, 90 [my emphasis].

14. Robin Scroggs, 'New Being: Renewed Mind: New Perception – Paul's View of the Source of Ethical Insight.' In *The Text and the Times: New Testament Essays for Today*. Minneapolis: Fortress Press, 1993, 167–83.

15. The same image is used in 1 Clement 36:2 and 59:3. This early Christian letter is thought to date from around the year 96.

16. For a full discussion, see Alister E. McGrath, 'The Privileging of Vision: Lewis's Metaphors of Light, Sun, and Sight'. In *The Intellectual World of C. S. Lewis*. Oxford: Wiley-Blackwell, 2013, 83–104.

17. C.S. Lewis, *The Pilgrim's Regress*. London: Geoffrey Bles, 1950, 176–7.

18. Stuart Feder, *Gustav Mahler: A Life in Crisis*. New Haven, CT: Yale University Press, 2004, 150–1.

19. John Alexander Mackay, *A Preface to Christian Theology*. London: Nisbet, 1942, 27–54.

20. For a biography, see John Mackay Metzger, *The Hand and the Road: The Life and Times of John A. Mackay*. Louisville, Kentucky: Westminster John Knox Press, 2010.

21. Mackay, *Preface to Christian Theology*, 29.
22. Mackay, *Preface to Christian Theology*, 36.
23. Mackay, *Preface to Christian Theology*, 32.
24. Mackay, *Preface to Christian Theology*, 37.
25. Mackay, *Preface to Christian Theology*, 44.
26. Thomas Nagel, *The View from Nowhere*. New York: Oxford University Press, 1986, 67–89.
27. Alwyn Ruddock, 'The Earliest Original English Seaman's Rutter and Pilot's Chart'. *Journal of Navigation* 14, no. 4 (1961): 409–43.
28. Hermann Minkowski, 'Raum und Zeit'. *Jahresbericht der deutschen Mathematiker-Vereinigung* 18 (1909): 75–88.
29. J. Daniel Gifford, *Robert A. Heinlein: A Reader's Companion*. Sacramento: Nitrosyncretic Press, 2000, 110–11.
30. Einstein, letter of 21 March 1955; Pierre Speziali, ed., *Albert Einstein – Michele Besso Correspondence, 1903–55*. Paris: Hermann, 1972, 537–8.
31. See P.A. Schilpp, ed., *The Philosophy of Rudolf Carnap*. La Salle, IL: Open Court Publishing, 1963, 37–8.
32. Lewis, *Essay Collection*, 438–43.
33. C.S. Lewis, *An Experiment in Criticism*. Cambridge: Cambridge University Press, 1992, 137.
34. Lewis, *An Experiment in Criticism*, 140–1.

Chapter 5

1. Jeanette Winterson, *Why Be Happy When You Could Be Normal?* London: Vintage, 2012, 68.
2. There is a massive literature. A good starting point is Michael J. MacKenzie and Roy F. Baumeister, 'Meaning in Life: Nature, Needs, and Myth'. In *Meaning in Positive and Existential Psychology*, ed. Alexander Batthyany and Pninit Russo-Netze. New York: Springer, 2014, 25–38.
3. Joanna Collicutt, *The Psychology of Christian Character Formation*. Norwich: SCM Press, 2015, 142–59.
4. Michael F. Steger, 'Meaning in Life'. In *Oxford Handbook of Positive Psychology*, ed. Shane J. Lopez. Oxford, UK: Oxford University Press, 2009, 679–87.
5. Steger, 'Meaning in Life', 682.
6. Paul T.P. Wong, ed., *The Human Quest for Meaning: Theories, Research, and Applications*. 2nd edn. New York: Routledge, 2012; Phillip R. Shaver and Mario Mikulincer, eds. *Meaning, Mortality, and*

Choice: The Social Psychology of Existential Concerns. Washington, DC: American Psychological Association, 2012.

7. Crystal L. Park and Ian A. Gutierrez, 'Global and Situational Meanings in the Context of Trauma: Relations with Psychological Well-Being'. *Counselling Psychology Quarterly* 26, no. 1 (2013): 8–25.

8. Robert A. Emmons, *The Psychology of Ultimate Concerns: Motivation and Spirituality in Personality*. New York: Guilford Press, 1999.

9. See the excellent account of this in Eric Hobsbawm, *How to Change the World: Marx and Marxism 1840–2011*. London: Little, Brown, 2012.

10. C. Daniel Batson and E.L. Stocks, 'Religion: Its Core Psychological Function'. In *Handbook of Experimental Existential Psychology*, ed. Jeff Greenberg, Sander L. Koole and Tom Pyszczynski. New York: Guilford, 2013, 141–55.

11. Ralph W. Hood, Peter C. Hill and W. Paul Williamson, *The Psychology of Religious Fundamentalism*. New York: Guilford Press, 2005; Ralph W. Hood, Peter C. Hill and Bernard Spilka, *The Psychology of Religion: An Empirical Approach*. 4th edn. New York: Guilford Press, 2015.

12. Joshua A. Hicks and Laura A. King, 'Meaning in Life and Seeing the Big Picture: Positive Affect and Global Focus'. *Cognition and Emotion* 21, no. 7 (2007): 1577–84.

13. Ludwig Wittgenstein, *Tractatus Logico-Philosophicus*, Abingdon: Routledge, 2001, 88.

14. Terrence W. Tilley, 'Problems of Theodicy: Background'. In *Physics and Cosmology: Scientific Perspectives on the Problem of Natural Evil*, ed. Nancey Murphy, Robert J. Russell and William R. Stoeger. Vatican City: Vatican Observatory, 2007, 35–51; quote at p. 45.

15. For the issues, see Alister E. McGrath, *Re-Imagining Nature: The Promise of a Christian Natural Theology*. Oxford: Wiley-Blackwell, 2016, 73–100.

16. For Traherne's own approach, see Belden C. Lane, 'Thomas Traherne and the Awakening of Want', *Anglican Theological Review* 81, no. 4 (1999): 651–64.

17. Thomas Traherne, *Centuries of Meditations*, New York: Cosimo, 2010, 20.

18. Robert Iliffe, 'Newton, God, and the Mathematics of the Two Books'. In *Mathematicians and Their Gods: Interactions between Mathematics and Religious Beliefs*, ed. Snezana Lawrence and Mark McCartney. Oxford: Oxford University Press, 2015, 121–44.

19. Richard Dawkins, *River out of Eden: A Darwinian View of Life*. London: Phoenix, 1995, 133.

20. Karl R. Popper, 'Natural Selection and the Emergence of Mind'. *Dialectica* 32 (1978): 339–55; quote at p. 342.

21. Alexander Rosenberg, *The Atheist's Guide to Reality: Enjoying Life without Illusions*. New York: W.W. Norton, 2011, 7–8.

22. Robert F. Almeder, *Harmless Naturalism: The Limits of Science and the Nature of Philosophy*. Chicago: Open Court, 1998, 36–8.

23. See the important argument of Scott James, *An Introduction to Evolutionary Ethics*. Oxford: Wiley Blackwell, 2011, 133–8.

24. Charles Dickens, *Hard Times*. London: Wordsworth, 1995, 3.

25. Dickens, *Hard Times*, 11.

26. Dickens, *Hard Times*, 75.

27. Ludwig Wittgenstein, *Notebooks, 1914–1916*. New York: Harper, 1961, 74.

28. Elinor Ochs and Lisa Capps, 'Narrating the Self'. *Annual Review of Anthropology* 25 (1996): 19–43.

29. Michael Bamberg, 'Who Am I? Narration and Its Contribution to Self and Identity'. *Theory & Psychology* 21, no. 1 (2011): 3–24.

30. Lewis P. Hinchman and Sandra K. Hinchman, eds, *Memory, Identity, Community: The Idea of Narrative in the Human Sciences*. Albany, NY: State University of New York Press, 1997.

31. Jonathan Gottschall, *The Storytelling Animal: How Stories Make Us Human*. Boston: Houghton Mifflin Harcourt, 2012, 87–116.

32. Joseph Henderson, 'Ancient Myth and Modern Man'. In Carl G. Jung, *Man and His Symbols*. New York: Doubleday, 1964, 104–57.

33. J.R.R. Tolkien, *Tree and Leaf*. London: HarperCollins, 2001, 56. For reflections on this theme, see Fabienne Claire Caland, 'Le mythos spermatikos'. In *Horizons du mythe*, ed. Denise Brassard and Fabienne Claire Caland. Montréal: Cahiers du CELAT, 2007, 7–32.

34. Tolkien, *Tree and Leaf*, 71. This theme is discussed further in Verlyn Flieger, *Splintered Light: Logos and Language in Tolkien's World*. Kent, OH: Kent State University, 2002.

35. Ibid.

Chapter 6

1. Thomas Nagel, *The Last Word*. Oxford: Oxford University Press, 1997, 130.

2. Carl Sagan, 'Why We Need To Understand Science'. *Skeptical Inquirer* 14, no. 3 (Spring 1990).

3. For the issues, see Hans J. Ladegaard, 'Demonising the Cultural Other: Legitimising Dehumanisation of Foreign Domestic Helpers in the Hong Kong Press'. *Discourse, Context & Media* 2, no. 3 (2013): 131–40.

4. Steven Pinker, *The Blank Slate: The Modern Denial of Human Nature*. New York: Viking, 2002, 105–20.

5. Pinker, *The Blank Slate*, 110–11.

6. Hermann Hesse, 'Die Sehnsucht unser Zeit nach einer Weltanschauung'. *Uhu* 2 (1926): 3–14.

7. See Jonathan Haidt, 'The Emotional Dog and Its Rational Tail: A Social Intuitionist Approach to Moral Judgment'. *Psychological Review* 108, no. 4 (2001): 814–34.

8. 'Parce que nous sommes au monde, nous sommes condamnés au sens'. Maurice Merleau-Ponty, *Phénoménologie de la perception*. Paris: Librairie Gallimard, 1945, xiv.

9. The view of Michael Oakeshott: see especially his *Rationalism in Politics, and Other Essays*. New York: Basic Books, 1962.

10. Oakeshott, *Rationalism in Politics*, 6.

11. Howard Margolis, *Patterns, Thinking, and Cognition: A Theory of Judgment*. Chicago: University of Chicago Press, 1987.

12. Margolis, *Patterns, Thinking, and Cognition*, 21.

13. Daniel C. Krawczyk, 'Contributions of the Prefrontal Cortex to the Neural Basis of Human Decision Making'. *Neuroscience and Biobehavioral Reviews* 26 (2002): 631–64; Jonathan Haidt, *The Righteous Mind: Why Good People Are Divided by Politics and Religion*. New York: Pantheon Books, 2012, 3–108.

14. See, for example, Herbert McCabe, *Faith within Reason*. London: Continuum, 2007. An older rebuttal of 'believing in God as an object' can be found in Emil Brunner, *Our Faith*. London: SCM Press, 1949, 13–14.

15. William Ralph Inge, *Faith and Its Psychology*. New York: Charles Scribner's Sons, 1910, 197.

16. For a philosophically rigorous account of Feuerbach's views, see Larry Johnston, *Between Transcendence and Nihilism: Species-Ontology in the Philosophy of Ludwig Feuerbach*, New York: Peter Lang, 1995.

17. Brendan Wallace, *Getting Darwin Wrong: Why Evolutionary Psychology Won't Work*. Exeter: Imprint Academic, 2010, 4.

18. Nagel, *The Last Word*, 130.

19. Haidt, 'The Emotional Dog and Its Rational Tail'.

20. For Lewis's wrestling with this issue, see Alister E. McGrath, *C. S. Lewis – A Life: Eccentric Genius, Reluctant Prophet*. London: Hodder & Stoughton, 2013, 135–41.
21. Michael J. Buckley, *At the Origins of Modern Atheism*. New Haven, CT: Yale University Press, 1987.
22. Augustine of Hippo, *Confessions*, I.i.1.
23. Alan D. Schrift, *Nietzsche and the Question of Interpretation: Between Hermeneutics and Deconstruction*. New York: Routledge, 1990, 144–68.
24. For the chaotic implications of this position, see Peter Stoicheff, 'The Chaos of Metafiction'. In *Chaos and Order: Complex Dynamics in Literature and Science*, ed. N. Katherine Hayles. Chicago: University of Chicago Press, 1991, 85–99.
25. Richard Rorty, *Consequences of Pragmatism*. Minneapolis, MN: University of Minneapolis Press, 1982, xlii.
26. Richard J. Bernstein, *Philosophical Profiles: Essays in a Pragmatic Mode*. Philadelphia: University of Pennsylvania Press, 1986, 53–4.
27. David Harris, *From Class Struggle to the Politics of Pleasure: The Effects of Gramscianism on Cultural Studies*. London: Routledge, 1992.
28. Philip E. Devine, *Natural Law Ethics*. Westport, CT: Greenwood Press, 2000, 32–4.
29. Alasdair C. MacIntyre, *After Virtue: A Study in Moral Theory*. 3rd edn. Notre Dame, IN: University of Notre Dame Press, 2007, 107.
30. Kathleen Wall, 'Ethics, Knowledge, and the Need for Beauty: Zadie Smith's *On Beauty* and Ian McEwan's *Saturday*'. *University of Toronto Quarterly* 77, no. 2 (2008): 757–88; quote at p. 757.
31. For an extended discussion of these issues, see Joseph Rouse, *Engaging Science: How to Understand Its Practices*. Ithaca, NY: Cornell University Press, 1996; idem, *Knowledge and Power: Towards a Political Philosophy of Science*. Ithaca: Cornell University Press, 1987.
32. Iris Murdoch, *The Sovereignty of Good*. London: Macmillan, 1970. For discussion and analysis, see Maria Antonaccio and William Schweiker, eds, *Iris Murdoch and the Search for Human Goodness*. Chicago: University of Chicago Press, 1996.
33. Murdoch, *The Sovereignty of Good*, 88.
34. Fyodor Dostoevsky, *Devils*, trans. Michael R. Katz. Oxford: Oxford University Press, 1992, 691. The Russian title of the novel can also be translated into English as *The Demons* or *The Possessed*.

35. Joseph Frank and David I. Goldstein, eds, *Selected Letters of Fyodor Dostoyevsky*, trans. Andrew R. MacAndrew. New Brunswick, NJ: Rutgers University Press, 1987, 446.

36. For an excellent discussion, see C. Stephen Evans, *Faith Beyond Reason: A Kierkegaardian Account*. Grand Rapids, MI: Eerdmans, 1998.

37. Susan R. Wolf, *Meaning in Life and Why It Matters*. Princeton, NJ: Princeton University Press, 2010, 7.

38. Wolf, *Meaning in Life and Why It Matters*, 10.

39. Wolf, *Meaning in Life and Why It Matters*, 11.

40. Wolf, *Meaning in Life and Why It Matters*, 10.

41. Wolf, *Meaning in Life and Why It Matters*, 11.

42. See especially C.S. Lewis, *The Abolition of Man*. London: Oxford University Press, 1943.

43. For the defence of this position, see Richard Rorty, *Achieving Our Country*. Cambridge, MA: Harvard University Press, 1999.

44. On Kierkegaard's important discussion of the 'subjectivity of truth', see Merigala Gabriel, *Subjectivity and Religious Truth in the Philosophy of Søren Kierkegaard*. Macon, GA: Mercer University Press, 2010.

45. Letter to William Temple, Archbishop of Canterbury, 7 September 1943; *The Letters of Dorothy L. Sayers: Volume II, 1937 to 1943*, ed. Barbara Reynolds. New York: St. Martin's Press, 1996, 429.

46. For detailed comment on this work, see Erasmo Leiva-Merikakis, *Love's Sacred Order: The Four Loves Revisited*. San Francisco: Ignatius Press, 2000.

47. C.S. Lewis, *The Four Loves*. New York: Harcourt Brace Jovanovich, 1960, 103.

48. See Lynne Viola, *Peasant Rebels Under Stalin: Collectivization and the Culture of Peasant Resistance*. Oxford University Press, 1996, 13–44.

49. Richard Sennett, *The Corrosion of Character: The Personal Consequences of Work in the New Capitalism*. New York: Norton, 1998.

50. Lewis, *The Four Loves*, 180.

Chapter 7

1. Anne Frank, diary entry for 15 July 1944; *Anne Frank: The Diary of a Young Girl*. New York: Bantam, 1993, 263.

2. John Locke, *Works*, 10 vols, London: Thomas Tegg, 1823, vol. 8, p. 447.

3. See especially Jonathan Haidt, *The Righteous Mind: Why Good People Are Divided by Politics and Religion*. New York: Pantheon Books, 2012.

4. See the analysis in Lee Braver, *Groundless Grounds: A Study of Wittgenstein and Heidegger*. Cambridge, MA: MIT Press, 2012.

5. Bertrand Russell, *A History of Western Philosophy*. London: George Allen & Unwin Ltd, 1946, xiv.

6. Isaiah Berlin, *Concepts and Categories: Philosophical Essays*. New York: Viking Press, 1979, 2–5; 161–2.

7. Berlin, *Concepts and Categories*, 114–15.

8. Blaise Pascal, *Pensées*. Mineola, NY: Dover Publications, 2003, 64.

9. Bertrand Russell, *A History of Western Philosophy*. London: George Allen & Unwin Ltd, 1946, xiv.

10. Friedrich Nietzsche, *Götzen-Dämmerung; oder Wie man mit dem Hammer philosophiert*. Munich: Hanser, 1954, 7. 'Hat man sein warum? des Lebens, so verträgt man sich fast mit jedem wie?' Nietzsche then added the teasing comment: 'Human beings don't strive for happiness (only the English do that).'

11. Viktor E. Frankl, *Man's Search for Meaning*. New York: Simon & Schuster, 1963.

12. Kenneth I. Pargament, Bruce W. Smith, Harold G. Koenig and Lisa Perez, 'Patterns of Positive and Negative Religious Coping with Major Life Stressors'. *Journal for the Scientific Study of Religion* 37, no. 4 (1998): 710–24. For discussion, see Joanna Collicutt, *The Psychology of Christian Character Formation*. Norwich: SCM Press, 2015, 232–3.

13. See Manfred Zaumseil, Silke Schwarz, Mechthild von Vacano, Gavin Brent Sullivan and Johana E. Prawitasari-Hadiyono, eds. *Cultural Psychology of Coping with Disasters: The Case of an Earthquake in Java, Indonesia*. New York: Springer, 2014, 53–4.

14. Arthur Koestler, *The Invisible Writing: An Autobiography*. Boston: Beacon Press, 1954, 13.

15. Joanna Collicutt McGrath, 'Post-Traumatic Growth and the Origins of Early Christianity'. *Mental Health, Religion and Culture* 9 (2006): 291–306.

16. Frank, *The Diary of a Young Girl*, 263.

17. For a useful account of Lewis's notion of rationality, see Alister E. McGrath, 'An Enhanced Vision of Rationality: C. S. Lewis on the Reasonableness of Christian Faith'. *Theology* 116, no. 6 (2013): 410–17.

18. C.S. Lewis, *The Problem of Pain*. HarperCollins: New York, 2001, 91. Note also the later comment (p. 94) that pain 'plants the flag of truth within the fortress of a rebel soul'.

19. Letter to Warnie Lewis, 3 December 1939; *Letters*, vol. 2, 302. Emphasis in original.

20. C.S. Lewis, *A Grief Observed*. San Francisco: HarperCollins, 2001, 6–7. For a perceptive analysis of the issues raised by this work, see Ann Loades, 'C. S. Lewis: Grief Observed, Rationality Abandoned, Faith Regained'. *Literature and Theology* 3 (1989): 107–21.

21. For the phenomenon, see Joanna Collicutt McGrath, 'Post-Traumatic Growth and the Origins of Early Christianity'. *Mental Health, Religion and Culture* 9 (2006): 291–306.

22. See, for example, Richard Swinburne, *The Existence of God*. 2nd edn. Oxford: Oxford University Press, 2004, 236–72.

23. For example, see James I. Packer, 'An Introduction to Systematic Spirituality'. *Crux* 26, no. 1 (1990): 2–8. A good account of the possibilities of enrichment and integration of theology and spirituality can be found in F. LeRon Shults and Steven J. Sandage, *Transforming Spirituality: Integrating Theology and Psychology*. Grand Rapids, MI: Baker Academic, 2006.

24. Ellen M. Ross, *The Grief of God: Images of the Suffering Jesus in Late Medieval England*. New York: Oxford University Press, 1997.

Chapter 8

1. David Brewster, *Life of Sir Isaac Newton*. New edn, rev. W.T. Lynn. London: Tegg, 1875, 303.

2. Loren Eiseley, *The Star Thrower*. New York: Harcourt Brace & Co., 1978, 267–79.

3. Eiseley, *The Star Thrower*, 278.

4. Eiseley, *The Star Thrower*, 271.

5. Eiseley, *The Star Thrower*, 169–85.

6. Eiseley, *The Star Thrower*, 182.

7. For a good account, see Robert C. Fuller, *Wonder: From Emotion to Spirituality*. Chapel Hill, NC: University of North Carolina Press, 2006. Note especially Fuller's reflections (pp. 69–79) on the role of wonder in the thought of William James.

8. Johanna Spyri, *Heidis Lehr- und Wanderjahre*. Villingen: Nexx Verlag, 2014, 45.

9. Anthony J. Carroll, 'Disenchantment, Rationality, and the Modernity of Max Weber'. *Forum Philosophicum* 16, no. 1 (2011): 117–37.

10. Philip Fisher, *Wonder, the Rainbow, and the Aesthetics of Rare Experiences*. Cambridge, MA: Harvard University Press, 1998.

11. For comment on these lines from Keats's poem *Lamia*, see Andrew Motion, *Keats*. Chicago: University of Chicago Press, 1998, 431–7; Fisher, *Wonder, the Rainbow, and the Aesthetics of Rare Experiences*, 87–93.

12. A good example is Richard Dawkins, *Unweaving the Rainbow: Science, Delusion and the Appetite for Wonder*. London: Penguin Books, 1998.

13. Huib J. Zuidervaart, 'The "True Inventor" of the Telescope. A Survey of 400 Years of Debate'. In *The Origins of the Telescope*, ed. Albert Van Helden, Sven Dupré, Rob van Gent and Huib Zuidervaart. Amsterdam: Koninklijke Nederlandse Akademie van Wetenschappen, 2010, 9–44.

14. Albert Van Helden, 'The Telescope in the Seventeenth Century'. *Isis* 65, no. 1 (1974): 38–58.

15. For the historical aspects of this insight, see Robin Downie, 'Science and the Imagination in the Age of Reason'. *Medical Humanities* 27 (2001): 58–63.

16. Richard P. Feynman, *The Character of Physical Law*. Boston: MIT Press, 1988, 127–8.

17. I here paraphrase the important analysis in Werner Heisenberg, *Die Ordnung der Wirklichkeit*. Munich: Piper Verlag, 1989, 38–52.

18. Heisenberg, *Die Ordnung der Wirklichkeit*, 44.

19. I explore some of these further in Alister E. McGrath, *Enriching our Vision of Reality: Theology and the Natural Sciences in Dialogue*. London: SPCK, 2016.

20. William James, *The Will to Believe*. New York: Dover Publications, 1956, 51.

21. For an accessible introduction to this phenomenon, see Ian Morison, *Introduction to Astronomy and Cosmology*. Oxford: Wiley, 2008, 1–3.

22. Heisenberg, *Die Ordnung der Wirklichkeit*, 44.

23. For Brunner's point, see Alister E. McGrath, *Emil Brunner: A Reappraisal*. Oxford: Wiley-Blackwell, 2016, 50–4.

24. For my own contributions to discussion of this topic, see especially Alister E. McGrath, *Re-Imagining Nature: The Promise of a Christian Natural Theology*. Oxford: Wiley-Blackwell, 2016.

25. Kathryn Schifferdecker, *Out of the Whirlwind: Creation Theology in the Book of Job.* Cambridge, MA: Harvard University Press, 2008.

26. For comment on the reception of these lines, see Hartmut Günther, 'Psalm 19 und die Verkündigung des Evangeliums unter den Völkern: Zum Verständnis von Psalm 19 bei Paulus und bei Luther'. In *Unter einem Christus sein und streiten: Festschrift für Friedrich Wilhelm Hopf*, ed. Jobst Schöne and Volker Stolle. Erlangen: Verlag ELM, 1980, 11–25.

27. James Barr, 'Do We Perceive the Speech of the Heavens? A Question in Psalm 19'. In *The Psalms and Other Studies on the Old Testament*, ed. Jack C. Knight and Lawrence A. Sinclair. Nashotah, WI: Nashotah House Seminary, 1990, 11–17; Randall C. Zachman, 'The Universe as the Living Image of God: Calvin's Doctrine of Creation Reconsidered'. *Concordia Theological Monthly* 61 (1997): 299–312.

28. McGrath, *Re-Imagining Nature*, 18–22.

29. For a fuller discussion of this theme, see Alister E. McGrath, *Dawkins' God: From the Selfish Gene to the God Delusion.* 2nd edn. Oxford: Wiley-Blackwell, 2015, 166–9.

30. Richard Dawkins, 'A Survival Machine'. In *The Third Culture*, ed. John Brockman. New York: Simon & Schuster, 1996, 75–95; quote at p. 85.

31. Eugene Wigner, 'The Unreasonable Effectiveness of Mathematics'. *Communications on Pure and Applied Mathematics* 13 (1960): 1–14, especially p. 8.

32. E.g., see John Polkinghorne, *Science and Creation: The Search for Understanding.* London: SPCK, 1988, 20–1. For further discussion, see Daniel J. Cohen, *Equations from God: Pure Mathematics and Victorian Faith.* Baltimore, MD: Johns Hopkins University Press, 2007.

33. Paul Dirac, 'The Relation between Mathematics and Physics'. *Proceedings of the Royal Society of Edinburgh* 59 (1938–9): 122–9.

34. Johann Kepler, *Gesammelte Werke.* Ed. Max Caspar, 22 vols, Munich: C.H. Beck, 1937–83, vol. 6, 233.

35. P.A. Schilpp, ed., *The Philosophy of Rudolf Carnap.* La Salle, IL: Open Court Publishing, 1963, 37–8.

Chapter 9

1. Blaise Pascal, *Pensées.* Mineola, NY: Dover Publications, 2003, 61.

2. Jacques Monod, *Chance and Necessity: An Essay on the Natural Philosophy of Modern Biology.* London: Penguin, 1997, 2.

3. *The Rubáiyát of Omar Khayyám*, trans. Edward Fitzgerald. London: Wordsworth, 1993, 66.
4. Marcel Arnould, Stephane Goriely and Kohji Takahashi, 'The R-Process of Stellar Nucleosynthesis: Astrophysics and Nuclear Physics Achievements and Mysteries'. *Physics Reports* 450 (2007): 97–213.
5. Gordon C.F. Bearn, *Waking to Wonder: Wittgenstein's Existential Investigations*. Albany, NY: State University of New York Press, 1997, 163–97.
6. Michael Mayne, *This Sunrise of Wonder: Letters for the Journey*. London: Darton, Longman and Todd, 2008, 15.
7. G.K. Chesterton, *Autobiography*. San Francisco: Ignatius, 2006, 99.
8. Chesterton, *Autobiography*, 134.
9. Stephen Hawking, *Black Holes: The Reith Lectures*. London: Bantam, 2006.
10. Gustave Flaubert, *Correspondance*, 5 vols, Paris: Gallimard, 1991, vol. 3, 220.
11. Heidegger's views on what constitutes 'authentic existence' are complex and contested. For a useful starting point, see Mark A. Wrathall and Jeff Malpas, eds, *Heidegger, Authenticity and Modernity: Essays in Honor of Hubert L. Dreyfus*. Cambridge, MA: MIT Press, 2000.
12. Douglas Burton-Christie, 'Place-Making as Contemplative Practice'. *Anglican Theological Review* 91, no. 3 (2009): 347–71.
13. Walter Brueggemann, *The Land: Place as Gift, Promise, and Challenge in Biblical Faith*. 2nd edn. Philadelphia: Fortress Press, 2002, 5. More generally, see Edward S. Casey, *The Fate of Place: A Philosophical History*. Berkeley, CA: University of California Press, 1998, 285–330.
14. Marc Augé, *Non-lieux: Introduction à une anthropologie de la surmodernité*. Paris: Éditions du Seuil, 1992. See further Emer O'Beirne, 'Mapping the *Non-Lieu* in Marc Augé's Writings'. *Forum for Modern Language Studies* 42/1 (2006): 38–50.
15. Matthew Arnold, *Poems*, 2 vols, London: Macmillan, 1891, vol. 2, 185.
16. C.S. Lewis, *Essay Collection*. London: HarperCollins, 2002, 96–106.
17. C.S. Lewis, *Mere Christianity*. London: HarperCollins, 2002, 136–7.
18. For examples of such worldviews, see Carole M. Cusack, *Invented Religions: Imagination, Fiction and Faith*. Farnham: Ashgate, 2010, especially 7–25.

19. For a good introduction, see Gérard Vallée, *The Shaping of Christianity: The History and Literature of Its Formative Centuries (100–800)*. New York: Paulist Press, 1999, 64–73.

20. G.K. Chesterton, *Tremendous Trifles*. London: Methuen, 1909, 209.

21. For a full discussion, see Erhardt Güttgemanns, *Der leidende Apostel und sein Herr; Studien zur paulinischen Christologie*. Göttingen: Vandenhoeck & Ruprecht, 1966. More generally, see Andrew T. Lincoln, *Paradise Now and Not Yet: Studies in the Role of the Heavenly Dimension in Paul's Thought with Special Reference to His Eschatology*. Cambridge: Cambridge University Press, 2004.

22. Cyprian of Carthage, *On Mortality*, 7.25.

23. For a detailed study, see Avi Faust, *Judah in the Neo-Babylonian Period: The Archaeology of Desolation*. Atlanta, GA: Society of Biblical Literature, 2012.

24. Ellen F. Davis, 'Singing for the Peace of Jerusalem: Songs of Zion in the Twenty-First Century'. In *The Bible and Spirituality: Exploratory Essays in Reading Scripture Spiritually*, ed. Andrew T. Lincoln, Gordon McConville and Lloyd K. Pietersen. Eugene, OR: Wipf & Stock, 2013, 75–94.

25. Joseph Pieper, *On Hope*. San Francisco: Ignatius Press, 1986, 38.

26. Edward Said, 'The Mind of Winter: Reflections on Life in Exile'. *Harper's Magazine*, September 1984, 49–55; quote at p. 49.

27. Simone Weil, *L'enracinement. Prélude à une déclaration des devoirs envers l'être humain*. Paris: Éditions Gallimard, 1949.

Chapter 10

1. Aleksandr Solzhenitsyn, *The Gulag Archipelago 1918–56*. London: Harvill Press, 2003, 75.

2. Charles Frankel, *The End of the Dinosaurs: Chicxulub Crater and Mass Extinctions*. Cambridge: Cambridge University Press, 1999; Michael J. Benton, *When Life Nearly Died: The Greatest Mass Extinction of All Time*. London: Thames & Hudson, 2003.

3. William Hazlitt, *Essays*. London: Walter Scott, 1889, 269.

4. Baretti's remark was recorded by Boswell, although it is often misattributed to Samuel Johnson. For the quote, see Frank Brady and Frederick A. Pottle, eds, *Boswell on the Grand Tour: Italy, Corsica and France, 1765–1766*. London: Heinemann, 1955, 281.

5. For detailed documentation of this point, see Jonathan Glover, *Humanity: A Moral History of the Twentieth Century*. London: Pimlico, 2001.

6. Glover, *Humanity*, 7.

7. See the superb analysis in Carol Tavris and Elliot Aronson, *Mistakes Were Made (But Not by Me): Why We Justify Foolish Beliefs, Bad Decisions, and Hurtful Acts*. Boston: Mariner Books, 2015.

8. G.R. Evans, *Augustine on Evil*. Cambridge: Cambridge University Press, 1982; Rowan Williams, *On Augustine*. London: Bloomsbury, 2016, 79–105.

9. Louis F. Fieser, 'The Synthesis of Vitamin K'. *Science* 91 (1940): 31–6.

10. For the story of Fieser's intimate association with this development, see Robert M. Neer, *Napalm: An American Biography*. Cambridge, MA: Harvard University Press, 2015, 5–44.

11. Described in Neer, *Napalm*, 60.

12. Interview on BBC *Newsnight* programme, 15 September 2009.

13. John Gray, *Straw Dogs: Thoughts on Humans and Other Animals*. London: Granta, 2002, 29.

14. Gray, *Straw Dogs*, 14.

15. Gray, *Straw Dogs*, 4.

16. Gray, *Straw Dogs*, 92.

17. For some good introductions to this grim theme, see Levon Chorbajian and George Shirinian, eds, *Studies in Comparative Genocide*. New York: St. Martin's Press, 1999; Omer Bartov, Anita Grossmann and Mary Nolan, eds. *Crimes of War: Guilt and Denial in the Twentieth Century*. New York: New Press, 2002; Manus I. Midlarsky, *The Killing Trap: Genocide in the Twentieth Century*. Cambridge: Cambridge University Press, 2005; Adam Jones, *Genocide: A Comprehensive Introduction*. London: Routledge, 2006.

18. R.G. Collingwood, *An Autobiography*. London: Oxford University Press, 1939, 79.

19. Dan Edelstein, *The Enlightenment: A Genealogy*. Chicago: University of Chicago Press, 2010, 17.

20. I explain more of this in Alister E. McGrath, *Inventing the Universe: Why We Can't Stop Talking About Science, Faith and God*. London: Hodder & Stoughton, 2015.

21. *Washington Post*, 27 February 2016.

22. Richard Crouter, *Reinhold Niebuhr on Politics, Religion, and Christian Faith*. New York: Oxford University Press, 2010, 101–2.

23. Some liberal Protestants unwisely melded some Darwinian values into their ethics in the name of 'progressive religion': see, for example, Gary J. Dorrien, *Social Ethics in the Making: Interpreting an American Tradition*. Malden, MA: Wiley-Blackwell, 2009, 69–73. Most, however, recognised this was a false turn.

24. Thomas H. Huxley, 'Evolution and Ethics'. In Thomas H. Huxley, *Evolution and Ethics and Other Essays*. London: Macmillan, 1905, 46–116.

25. Huxley, 'Evolution and Ethics', 51–2.

26. Huxley, 'Evolution and Ethics', 53.

27. Huxley, 'Evolution and Ethics', 81–2.

28. Huxley, 'Evolution and Ethics', 81.

29. James D.G. Dunn, *The Theology of Paul the Apostle*. Grand Rapids, MI: Eerdmans, 1998, 51–78.

30. Alister E. McGrath, *Iustitia Dei: A History of the Christian Doctrine of Justification*. 3rd edn. Cambridge: Cambridge University Press, 2005, 197–206.

31. Sigmund Freud, *Civilization and its Discontents* in *Civilization, Society and Religion*, 20 vols, London: Penguin, 1991, vol. 12, 302.

32. Charles Darwin, *The Variation of Animals and Plants Under Domestication*, 2 vols, London: John Murray, 1868, vol. 1, 5–6.

33. Patricia A. Williams, 'Sociobiology and Original Sin'. *Zygon* 35, no. 4 (2000): 783–812.

34. Steve Jones, *In the Blood: God, Genes and Destiny*. London: HarperCollins, 1996, 207–42.

35. For an excellent study, see Alan Jacobs, *Original Sin: A Cultural History*. New York: HarperOne, 2008. For a thoughtful theological defence and exploration of this important concept, see Tatha Wiley, *Original Sin: Origins, Developments, Contemporary Meaning*. New York: Paulist Press, 2002.

36. Marjorie Suchocki, *The Fall to Violence*. New York: Continuum, 1994, 85.

37. A point stressed by Susanne Kappeler, *The Will to Violence: The Politics of Personal Behaviour*. Cambridge: Polity Press, 1995, 9.

38. Anne Frank, diary entry for 15 July 1944; *Anne Frank: The Diary of a Young Girl*. New York: Bantam, 1993, 263.

39. A good example of this is found in the conversion of Charles Colson, Richard Nixon's hatchet man and master of dirty tricks. Following his conviction in the aftermath of the Watergate scandal, Colson

recognised Lewis's description of sin in himself, and converted to Christianity: see Jonathan Aitken, *Charles W. Colson: A Life Redeemed*. London: Continuum, 2005, 192–211.

40. Steven J. Bartlett, *The Pathology of Man: A Study of Human Evil*. Springfield, IL: Thomas, 2005, 75–90.

41. Slavoj Žižek, *On Belief*. London: Routledge, 2001, 38. See the full discussion in Hannah Arendt, *The Origins of Totalitarianism*. New York: Meridian, 1958.

Chapter 11

1. Giovanni Pico della Mirandola, *Oration on the Dignity of Man*. Trans. Robert Caponigri. Chicago: Gateway Editions, 1956, 11.

2. Charles Darwin, *The Descent of Man*, 2 vols, London: Murray, 1871, vol. 2, 404–5.

3. See further Alister E. McGrath, *The Intellectual Origins of the European Reformation*. 2nd edn. Oxford: Blackwell, 2003.

4. As noted in the ground-breaking article by Paul O. Kristeller, 'The Myth of Renaissance Atheism and the French Tradition of Free Thought'. *Journal of the History of Philosophy* 6 (1968): 233–43.

5. See the classic study of Hanna H. Gray, 'Renaissance Humanism: The Pursuit of Eloquence'. *Journal of the History of Ideas* 24, no. 4 (1963): 497–514.

6. For a representative collection of his writings, see Paul Oskar Kristeller, *Renaissance Thought and the Arts: Collected Essays*. Princeton, NJ: Princeton University Press, 1990. For an assessment of his approach, see John Monfasani, ed., *Kristeller Reconsidered: Essays on His Life and Scholarship*. New York: Italica Press, 2006.

7. See, for example, Paul Kurtz, *What is Secular Humanism?* Amherst, NY: Prometheus Books, 2006.

8. For an excellent account, see Mason Olds, *American Religious Humanism*. Minneapolis, MN: University Press of America, 1996.

9. For example, see Mark Oppenheimer, 'Closer Look at Rift Between Humanists Reveals Deeper Divisions'. *New York Times*, 1 October 2010.

10. Ludwig Wittgenstein, *Culture and Value*. 2nd edn. Oxford: Blackwell, 1980, 44.

11. Giovanni Pico della Mirandola, *Oration on the Dignity of Man*, 11.

12. John J. Coughlin, 'Pope John Paul II and the Dignity of the Human Being'. *Harvard Journal of Law & Public Policy* 67 (2003): 65–79.

For the development of John Paul II's Christian humanism, see George Weigel, *Witness to Hope: The Biography of John Paul II*. New York: HarperCollins, 2005, 145–80.

13. Bernard Williams, *Morality: An Introduction to Ethics*. Cambridge: Cambridge University Press, 1993, 80.

14. Dawkins, *God Delusion*, 31.

15. Christopher Hitchens, *God Is Not Great: How Religion Poisons Everything*. New York: Twelve, 2007, 8.

16. Hitchens, *God is Not Great*, 10.

17. Daniel C. Dennett, 'The Bright Stuff', *New York Times*, 12 July 2003; Richard Dawkins, 'The Future Looks Bright' *Guardian*, 21 June 2003.

18. Hitchens, *God Is Not Great*, 5.

19. William Temple, *Nature, Man and God*. London: Macmillan, 1934, 22.

20. Terry Eagleton, *Hope Without Optimism*. New Haven, CT: Yale University Press, 2015.

21. See especially Reinhold Niebuhr, *Moral Man and Immoral Society: A Study in Ethics and Politics*. New York: Charles Scribner's Sons, 1932.

22. John W. Dodds, 'The Place of the Humanities in a World of War'. *Vital Speeches of the Day* 9 (1943): 311–14.

23. Mark Roseman, *The Villa, the Lake, the Meeting: Wannsee and the Final Solution*. London: Penguin Books, 2003.

24. For the document, see http://www.ghwk.de/ghwk/deut/protokoll.pdf. Note especially p. 8.

25. These were later published as *Mere Christianity* (1952). For details, see Alister E. McGrath, *C. S. Lewis – a Life: Eccentric Genius, Reluctant Prophet*. London: Hodder & Stoughton, 2013, 205–13.

26. Mahmood Mamdani, *When Victims Become Killers: Colonialism, Nativism, and the Genocide in Rwanda*. Princeton, NJ: Princeton University Press, 2002.

27. Pius XII, *Mystici Corporis Christi*, 94; http://w2.vatican.va/content/pius-xii/en/encyclicals/documents/hf_p-xii_enc_29061943_mystici-corporis-christi.html. I have altered the translation at several points to catch the sense of the original Latin.

28. For a good exploration of this idea and its relational implications, see Stanley J. Grenz, *The Social God and the Relational Self: A Trinitarian Theology of the Imago Dei*. Louisville, KY: Westminster John Knox Press, 2001.

29. Lactantius, *Divine Institutions* VI, 10.

30. Yechiel Michael Barilan, *Human Dignity, Human Rights, and Responsibility: The New Language of Global Bioethics and Biolaw*. Cambridge, MA: MIT Press, 2012, 57–64.

Chapter 12

1. Benjamin Disraeli, *Tancred; or, The New Crusade*. New York: Walter Dunne, 1904, 78. For the historical context, see Paul Smith, *Disraeli: A Brief Life*. Cambridge: Cambridge University Press, 1996, 86–8.

2. See the points made by Bastiaan T. Rutjens, Frenk van Harreveld and Joop van der Pligt, 'Yes We Can: Belief in Progress as Compensatory Control'. *Social Psychological and Personality Science* 1, no. 3 (2010): 246–52; Paul G. Bain et al., 'Collective Futures: How Projections About the Future of Society Are Related to Actions and Attitudes Supporting Social Change'. *Personality and Social Psychology Bulletin* 39 (2013): 523–39.

3. I here follow some basic themes outlined in Daniel Sarewitz, 'The Idea of Progress'. In *A Companion to the Philosophy of Technology*, ed. Jan Kyrre Berg Olsen Friis, Stig Andur Pedersen and Vincent F. Hendricks. Chichester, UK: Wiley-Blackwell, 2013, 303–7; Philip Kitcher, 'On Progress'. In *Performance and Progress: Essays on Capitalism, Business, and Society*, ed. Subramanian Rangan. Oxford: Oxford University Press, 2015.

4. 'The French Revolution as It Appeared to Enthusiasts at Its Commencement'. In William Wordsworth, *Collected Poems*. Ware: Wordsworth Editions, 2006, 245.

5. Frank Martela, 'Fallible Inquiry with Ethical Ends-in-View: A Pragmatist Philosophy of Science for Organizational Research'. *Organization Studies* 36, no. 4 (2015): 537–63.

6. See the comments of Anne McClintock, 'The Angel of Progress: Pitfalls of the Term "Post-Colonialism"'. *Social Text*, no. 31/32 (1992): 84–98.

7. Thomas Sowell, *On Classical Economics*. New Haven, CT: Yale University Press, 2006, 184.

8. Neil Harding, *Leninism*. London: Macmillan, 1996, 155.

9. See the classic critique of Marxism developed by Karl R. Popper, *The Poverty of Historicism*. London: Routledge & Kegan Paul, 1957.

10. Reinhold Niebuhr, *The Nature and Destiny of Man: A Christian Interpretation*, 2 vols, London: Nisbet, 1941, vol. 2, 248.

11. C.S. Lewis, *Mere Christianinity*. London: HarperCollins, 2002, 28–9.

12. The works of the philosopher John Gray should be noted here: see especially his *Straw Dogs: Thoughts on Humans and Other Animals*. London: Granta, 2003; and *The Silence of Animals: On Progress and Other Modern Myths*. London: Penguin Books, 2014.

13. Terry Eagleton, *Reason, Faith, and Revolution: Reflections on the God Debate*. New Haven, CT: Yale University Press, 2009. It is instructive to read Eagleton alongside the older work of Georges Sorel, *Les illusions du progres*. 3rd edn. Paris: Rivière, 1921.

14. Eagleton, *Reason, Faith, and Revolution*, 28.

15. Eagleton, *Reason, Faith, and Revolution*, 87–9.

16. Eugene Lyons, *Assignment in Utopia*. New Brunswick, NJ: Transaction Publishers, 1991, 280.

17. Terry Eagleton, *Hope Without Optimism*. New Haven, CT: Yale University Press, 2015, 11.

18. Eagleton, *Hope Without Optimism*, 9.

19. A point made forcefully in Michael J. Buckley, *At the Origins of Modern Atheism*. New Haven, CT: Yale University Press, 1987.

20. Stephen Gaukroger, *Francis Bacon and the Transformation of Early Modern Philosophy*. Cambridge: Cambridge University Press, 2001.

21. One of the best accounts of this technological transformation of warfare is found in Jonathan Glover, *Humanity: A Moral History of the Twentieth Century*. 2nd edn. New Haven, CT: Yale University Press, 2012. For Tolkien's experience of the First World War, see John Garth, *Tolkien and the Great War: The Threshold of Middle-Earth*. London: HarperCollins, 2004.

22. Alan Jacobs, 'Fall, Mortality, and the Machine: Tolkien and Technology'. *Atlantic Monthly*, 27 July 2012.

23. Michael Ruse, 'Charles Darwin and Artificial Selection'. *Journal of the History of Ideas* 36, no. 2 (1975): 339–50.

24. Paul Cartledge, *Sparta and Lakonia: A Regional History 1300 to 362 BC*. 2nd edn. New York: Routledge, 2002, 84.

25. Charles Darwin, *The Descent of Man*, 2 vols, London: John Murray, 1871, vol. 1, 168.

26. David J. Galton and Clare J. Galton, 'Francis Galton and Eugenics Today'. *Journal of Medical Ethics* 24 (1998): 99–105.

27. Mark B. Adams, Garland E. Allen and Sheila F. Weiss, 'Human Heredity and Politics: A Comparative Institutional Study of the Eugenics Record Office at Cold Spring Harbor (United States), the Kaiser Wilhelm Institute for Anthropology, Human Heredity, and

Eugenics (Germany), and the Maxim Gorky Medical Genetics Institute (USSR)'. *Osiris* 20 (2005): 232–62.

28. Marie Carmichael Stopes, *Radiant Motherhood: A Book for Those Who are Creating the Future*. London: Putnam's Sons, 1920, 223.

29. Mark B. Adams, 'The Politics of Human Heredity in the USSR, 1920–1940'. *Genome* 31, no. 2 (1989): 879–84.

30. For a good historical example, see Edwin Black, *War against the Weak: Eugenics and America's Campaign to Create a Master Race*. New York: Dialogue Press, 2012.

31. http://www.nickbostrom.com/ethics/values.pdf

32. Julian Savulescu, R.H.J. ter Meulen and Guy Kahane, *Enhancing Human Capacities*. Oxford: Wiley-Blackwell, 2011.

33. http://www.nickbostrom.com/papers/future.pdf

34. Nick Bostrom and Anders Sandberg, 'The Wisdom of Nature: An Evolutionary Heuristic for Human Enhancement'. In *Human Enhancement*, ed. Julian Savulescu and Nick Bostrom. Oxford: Oxford University Press, 2008, 375–416.

35. http://www.nickbostrom.com/papers/future.pdf

36. Victor C. Ferkiss, *Technological Man: The Myth and the Reality*. New York: New American Library, 1970, 34.

37. Ingmar Persson and Julian Savulescu, *Unfit for the Future: The Need for Moral Enhancement*. Oxford: Oxford University Press, 2012.

38. For example, see Nick Bostrom and Milan Cirkovic, eds, *Global Catastrophic Risks*. Oxford: Oxford University Press, 2007.

39. Eagleton, *Hope Without Optimism*, 31.

40. John Gray, *The Immortalization Commission: The Strange Quest to Cheat Death*. New York: Farrar, Straus and Giroux, 2011, 235.

41. Peter Berger, *The Sacred Canopy*. New York: Doubleday, 1965.

42. J.R.R. Tolkien, *Tree and Leaf*. London: HarperCollins, 2001, 87.

43. For what follows, see Gray, *Immortalization Commission*, 156–67.

44. Anna Dickinson, 'Quantifying Religious Oppression: Russian Orthodox Church Closures and Repression of Priests 1917–41'. *Religion, State & Society* 28 (2000): 327–35. See further William Husband, 'Soviet Atheism and Russian Orthodox Strategies of Resistance, 1917–1932'. *Journal of Modern History* 70 (1998): 74–107.

45. Also known as the 'League of the Militant Godless'. For its history and methods, see Daniel Peris, *Storming the Heavens: The Soviet League of the Militant Godless*. Ithaca, NY: Cornell University Press, 1998.

Chapter 13

1. John Banville, *The Sea*. New York: Vintage Books, 2005, 71.
2. Richard Dawkins, *A Devil's Chaplain: Selected Essays*. London: Weidenfeld & Nicolson, 2003, 19.
3. John Polkinghorne, *Theology in the Context of Science*. New Haven, CT: Yale University Press, 2008, 126.
4. Bertrand Russell, *A History of Western Philosophy*. London: George Allen & Unwin Ltd, 1946, xiv.
5. On Berlin's 'value pluralism', see John Gray, *Isaiah Berlin: An Interpretation of His Thought*. Princeton, NJ: Princeton University Press, 2013, 74–110; Mark Thompson, 'Versions of Pluralism: William Empson, Isaiah Berlin, and the Cold War'. *Literary Imagination* 8, no. 1 (2006): 65–87.
6. Gray, *Isaiah Berlin*, 109.
7. 'The Tables Turned'. In William Wordsworth, *Collected Poems*. Ware: Wordsworth Editions, 2006, 574.
8. Dacher Keltner and Jonathan Haidt, 'Approaching Awe, a Moral, Spiritual and Aesthetic Emotion'. *Cognition and Emotion* 17 (2003): 297–314; Patty van Cappellen and Vassilis Saroglou, 'Awe Activates Religious and Spiritual Feelings and Behavioral Intentions'. *Psychology of Religion and Spirituality* 4 (2012): 223–36.
9. Dante, *Inferno*, I.1–3.
10. Barbara Reynolds, in *The Divine Comedy*, trans. D.L. Sayers and B. Reynolds, 3 vols, London: Penguin Classics, 1986, vol. 3, 16. Reynolds' best study of Dante is *Dante: The Poet, the Political Thinker, the Man*. London: I.B.Tauris, 2006.

Index

Index

Do you wish this wasn't the end?
Are you hungry for more great teaching, inspiring
testimonies, ideas to challenge your faith?

Join us at www.hodderfaith.com, follow us on Twitter
or find us on Facebook to make sure you get the latest from
your favourite authors.

Including interviews, videos, articles, competitions
and opportunities to tell us just what you thought about
our latest releases.

www.hodderfaith.com

 HodderFaith

 @HodderFaith

 HodderFaithVideo

HODDER
WHERE FAITH IS INSPIRED